E. Pretsch, G. Tóth, M. E. Munk, M. Badertscher

Computer-Aided Structure Elucidation

E. Pretsch, G. Tóth, M. E. Munk,
M. Badertscher

Computer-Aided Structure Elucidation

Spectra Interpretation and
Structure Generation

Prof. Dr. Ernö Pretsch
Laboratorium für Organische Chemie
ETH Hönggerberg
HCI E313
CH-8093 Zürich
Switzerland

Prof. Dr. Morton E. Munk
Arizona State University
Department of Chemistry
Tempe, AZ 85287-1604
USA

Prof. Dr. Gábor Tóth
Technical University Budapest
Szt. Gellért tér 4
H-1111 Budapest
Hungary

Dr. Martin Badertscher
Laboratorium für Organische Chemie
ETH Hönggerberg
HCI G339
CH-8093 Zürich
Switzerland

Library of Congress Card No.: applied for

A catalogue record for this book is available from the British Library.

Bibliographische Information Der Deutschen Bibliothek
Die Deutsche Bibliothek verzeichnet diese Publikation in der Deutschen Nationalbiografie; detaillierte bibliografische Daten sind im Internet über http://dnb.ddb.de abrufbar.

A catalogue record for this publication is available from Die Deutsche Bibliothek

ISBN 3-527-30640-4

Preface

Spectroscopic techniques are particularly powerful in the structure elucidation of organic compounds when several methods are combined. However, an integral view of the associated rules, reference data, and tools is necessary in order to take full advantage of the possible synergies. This volume stresses the combined application of spectroscopic methods and puts special emphasis on two more recent developments: (1) the routine use of various two-dimensional NMR techniques and (2) the availability of computer programs for generating topological isomers on the basis of structural information. A set of 18 prototype examples of organic compounds are given to introduce the reader to the technique of combined spectral interpretation. In each case, some aspects of the use of the structure generator, Assemble 2.1, are presented. The program is enclosed on a CD and the reader is invited to make his on-hand experience.

The volume is not an introductory textbook and supposes the reader to have basic knowledge of the various spectroscopic methods (a selection of recommended reference books is given below). It is more intended for undergraduate students and technicians who want to gain experience in the combined application of spectroscopic methods. It will also be useful to specialists in other fields and non-chemists who want to get acquainted with the modern approach to structure elucidation. Finally, experts interested in learning about the possibilities provided by structure generators will also profit from this book.

Our special thanks go to Dr. József Kovács from the Budapest University of Technology and Economics for running and editing all NMR spectra and to Dr. Dorothée Wegmann for her expertise in eliminating many errors and inconsistencies. We also thank Upstream Solutions (CH–6052 Hergiswil, Switzerland) for providing free versions of the computer programs on the enclosed compact disk.

How to Install the Programs

The enclosed CD-ROM contains the computer programs Assemble 2.1.1 for Windows 3.1 upwards and NMR Prediction for Windows and Macintosh. A Macintosh version of Assemble 2.1.1 is currently being prepared. The programs are restricted to the problems of this book and other structures with ≤ 15 non-hydrogen atoms. Please, contact Upstream Solutions (www.upstream.ch, e-mail: info@upstream.ch) for the full versions. For system requirements and installation instructions, see file readme.txt on the CD-ROM.

Selected References

General introductions and excercises

L. D. Field, S. Sternhell, J. R. Kalman, Organic Structures from Spectra, 2nd Edition, Wiley, New York, 1995.

R. M. Silverstein, F. X. Webster, Spectrometric Identification of Organic Compounds, Wiley, New York, 1998.

M. Hesse, H. Meier, B. Zeeh, Spectroscopic Methods in Organic Chemistry, Thieme, Stuttgart, 1997.

2D NMR

T. D. W. Claridge, High-Resolution NMR Techniques in Organic Chemistry, Pergamon, Amsterdam, 1999.

J. K. M. Sanders, B. K. Hunter, Modern NMR Spectroscopy, 2nd ed., Oxford University Press, Oxford, 1993.

Reference data

E. Pretsch, P. Bühlmann, C. Affolter, Structure Determination of Organic Compounds, 3rd ed., Springer, Berlin, 2000.

Contents

1 Introduction to Computer-Aided Structure Elucidation

1.1 The Need for Structure Elucidation

The elucidation of the molecular structure of organic compounds is of widespread importance in the chemical and health-related sciences. Consider the case of taxol, one of the more recently discovered, promising drugs in the battle against cancer.

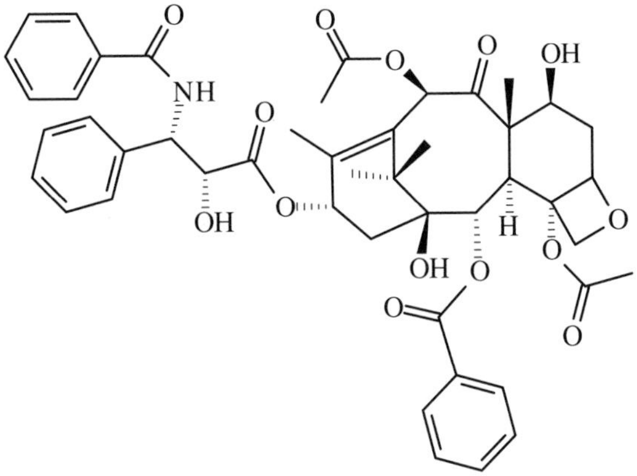

Fig. 1.1: Structural formula of taxol

This chemotherapeutic agent is a natural product. It was isolated from the bark of the yew, an evergreen tree native to Canada and the Northwest United States. The compound is present in only small amounts in the bark and the demand for it cannot, and should not, be met by stripping the bark of all yew trees. A practical, cost-effective, large-scale manufacturing process is required. However, the design of such a synthetic approach requires a detailed knowledge of the chemical structure; structure elucidation must precede synthetic design. Given the promise of the compound, minimizing the time required for structure elucidation is an important consideration.

Structure elucidation is important in the pharmaceutical industry in other ways as well. In the manufacture of pharmaceutical drugs, trace amounts of compounds that are by-products of the synthetic procedures used must be isolated, chemically identified, and determined to be harmless in the concentration in which they are present. Furthermore, as part of the process of bringing a drug to market, it may be necessary to determine the structure of the products of the metabolism of the drug in living systems. Structure elucidation is important to other areas of science as well – for example, environmental science and geology – and the availability of tools for this purpose is a necessary part of the research and development that is carried out on a daily basis in the laboratories of the world.

1.2 Types of Structure Elucidation Problems

Structure elucidation problems are usually of one of two types. In one type, generally referred to as structure verification, there is enough information available – perhaps based on the use of synthetic reactions whose outcomes are well established – for the chemist to propose a probable structure for the compound. However, verification of the suggested structure is required. In the second type, the information available to the chemist is insufficient to permit a structure to be proposed; the structure of the compound is unknown. The topics covered in this volume focus mainly on the second of these structure elucidation problems. Compounds of unknown structure derived from natural sources represent a large class of such structure elucidation problems.

1.3 The Basis of Structure Elucidation

One of the fundamental guiding principles of the chemical sciences is that the properties of a compound are a function of its molecular structure. This relationship can be expressed as:

Properties $= f$ (Structure)

Part of chemical research is devoted to elaborating the nature of the function f for a broad range of properties, e.g., NMR chemical shift, reaction rate in a particular reaction type, pK_a, and drug efficacy. The goal is to be able to predict a particular property of a compound based on its chemical structure, rather than determine it experimentally. The function f can be numerical, for example, a mathematical equation or a series of equations based on physical laws, non-numerical, for example, a series of steps defining a procedure (a heuristic procedure, or some combination of the two.

The underlying basis of structure elucidation derives from the inverse of the above relationship:

Structure $= f^{-1}$ (Properties)

Elaboration of the function f^{-1} for a given set of properties, for example, the collective spectroscopic data derived from a compound, could provide an approach to the determination of its structure. The function f^{-1} can also be numerical, or non-numerical in nature, or some combination of the two.

1.4   Historical Background

The problem of determining the chemical structure of compounds classed as "organic" is as old as organic chemistry itself. Organic compounds have been known since earliest antiquity and were originally believed to be found only in living organisms. However, this idea was shown to be invalid in 1828 when Friedrich Wöhler prepared urea, a compound known to be "organic" at the time, from ammonium cyanate, an inorganic compound. Today, it is generally agreed that organic chemistry is the study of compounds of carbon, regardless of their origin.

Many of the most distinguished early chemists struggled to understand the nature of the chemical structure of organic compounds and to develop methods for the determination of structure. However, it was not until 1859 that the stage was set for a major breakthrough. In that year, the German chemist, August Kekulé, in an intuitive rationalization of chemical behavior known at the time, first described the "structure" of simple carbon compounds in terms of a graphical representation of the connectivity and bonding of the atoms in the molecule. With modification, that graphical representation has survived to this day.

Building on the idea of the valence of atoms commonly found in organic compounds, Kekulé was able to rationalize the existence of different compounds of the same empirical formula (known as constitutional or structural isomers) in structural terms. He applied this reasoning to develop what is likely the earliest approach to structure elucidation. The process began with elemental analysis to determine the empirical formula of the unknown compound, which describes the kinds and ratios of the elements present in a molecule of the compound. Next, using the Kekulé graphical representation, all theoretically possible constitutional isomers were constructed. The assignment of the correct structure was then reduced to a problem of distinguishing among the different constitutional isomers. At the time, a solution to the problem required a study of the chemical behavior of the unknown and/or a synthesis of one or more of the isomers and comparison to the unknown. Of course, this approach to structure elucidation was limited to the simplest of compounds because of the paucity of chemical information available at the time. However, in some respects, the resemblance of Kekuké's basic approach to modern procedures in structure elucidation is striking.

From the beginning, the procedures used in structure elucidation were empirical in nature and based on observation and experience. For a long time, observed relationships between chemical behavior and structural features formed the foundation of the process. By the middle of the twentieth century, a new and powerful probe of chemical structure was beginning to emerge, molecular spectroscopy. Developments in this area came rapidly and led to a substantial and still growing knowledge base of correlations between spectral features and structural features. As a result, conventional structure elucidation became more heavily dependent on spectral data than on chemical behavior. Today, mass (MS), infrared (IR), nuclear magnetic resonance (NMR), and ultraviolet/visible (UV/vis) spectroscopies are powerful probes of molecular structure and the tools of choice in structure elucidation. The more recent availability of large, computer-readable spectral databases has significantly enhanced the importance of molecular spectroscopy in structure elucidation. Another comparative advantage of spectral data is the speed of their acquisition relative to chemical data.

1.5   The Nature of Structure Elucidation

Although no two chemists practice the "art" of structure elucidation in exactly the same way – and no two structure problems are exactly alike – generally three major stages can be discerned. The first stage is the *interpretation* of the collective spectroscopic data derived from the unknown; that is, the reduction of those data to structural information. (The same data may also lead to plausible molecular formulas for the unknown.) Some of the structural information is expressed as *substructures*, e.g., the presence or absence of a carbonyl group (C=O), and some is expressed as more *general structural features*, for example, the presence of one six-membered ring or the absence of multiple bonds.

The approach used in interpretation of the spectroscopic data depends in part on the depth of knowledge and experience of the interpreter. One may begin with any one of the spectroscopic methods and infer as many substructures and structural features as possible. Then, drawing on data from other spectroscopic methods, the initial information content is enriched and/or confirmed. Alternatively, one can follow the "creaming method" recommended here: Pick the most easily accessible and informative features from all of the spectra and infer from those spectral features only that structural information – substructures and structural features – which can be assigned with a high degree of reliability. Of course, the type of structure studied and the extent to which structural features are reflected in the different spectroscopic data will always play a decisive role.

Examples of the kinds of structural information that typically can be inferred from the various types of spectra are described below.

Mass spectrometry
- tentative assignment of molecular mass
- occurrence of elements from the nominal mass and isotope distribution
- type of compound from ion series and intensity distribution
- structural elements from neutral losses (mass differences of the first fragments relative to the molecular ion)

Infrared spectroscopy
- identification of functional groups from typical absorbances (OH, NH, special types of carbon-hydrogen bonds, groups with triple bonds and cumulated double bonds, C=O, C=C, aromatics). Except in the most simple cases, only very intense signals in the fingerprint region (1500–1000 cm^{-1}) should be interpreted at this stage.

^{1}H NMR spectroscopy
- total number of hydrogens from integrated intensities
- chemical environment of hydrogen atoms from their chemical shifts
- neighboring groups from the first-order splittings

- symmetrical higher-order spin systems

^{13}C NMR spectroscopy
- number of signals; possible minimal number of carbon atoms
- number of hydrogens attached to carbon atoms
- equivalent carbons by comparison of the integrals in the proton NMR spectrum and the presence of X–H (X ≠ C) groups from the IR spectrum
- chemical environment of carbon atoms from their chemical shifts

UV/vis spectroscopy
- identification of delocalized π-electron systems. Today, UV/vis spectroscopy has only a marginal relevance in structure elucidation work. For this reason, it is only used in a few cases and only tabulated data are given.

The second stage of structure elucidation is *structure generation*, that is, constructing *all* molecular structures compatible with the molecular formula of the unknown, as well as the inferred *substructures* and *general structural features* derived in the interpretation stage. Structure generation is a *bond-making* process. Although the exact procedure used by individual chemists may vary, there are some common features. The process is stepwise and involves joining the substructures together at their available bonding sites. There are generally many pathways by which this can take place. In the process fewer, but larger substructures are generated. At some stage, new constructions must be shown to be consistent with valence requirements of the elements, bond multiplicity at the bonding sites, substructures known to be absent, and the inferred general structural features, or they are rejected and a new pathway of bond-making is explored. In principle, the process is not complete until all possible pathways of bond-making are explored, thus insuring the generation of all complete molecules compatible with all of the inferred information.

In the hands of the chemist, this manual process of structure generation can be tedious and prone to error, in particular, the failure to consider all possible ways of joining bonding sites together. As a consequence, one or more of the plausible molecular structures may be overlooked. Thus, identifying *one* compound compatible with all of the available evidence does not ensure the correct assignment of the molecular structure of the unknown since there may be additional molecular structures equally compatible with the evidence. Therefore, great care is necessary in the structure generation step to ensure that the process is *exhaustive*.

In the event that more than one plausible molecular structure is generated, which is commonly the case, a third stage is needed that involves spectrum prediction and comparison. The spectral characteristics of each compatible molecular structure are predicted and then compared to the observed spectra of the unknown. Each compound that reveals little difference between predicted and observed spectroscopic data represents a plausible solution to the problem. If no compound among the candidates gives rise to a close fit between predicted and observed spectroscopic characteristics, the possibility of an inexhaustive structure generation and/or misinterpretation of the observed spectroscopic data must be considered. The overall process of structure elucidation is schematically described in Fig. 1.2.

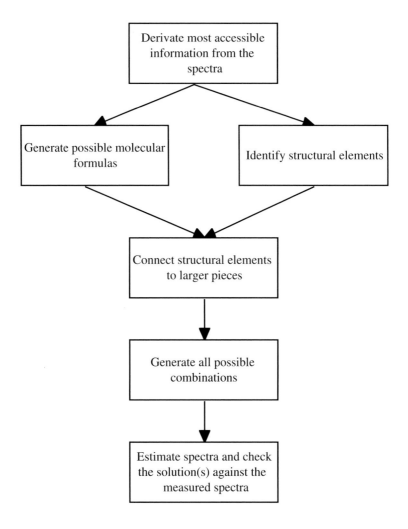

Fig. 1.2: General procedure of spectroscopic structure elucidation

1.6 The Role of the Computer in Structure Elucidation

Experienced chemists doing structure elucidation by conventional methods – that is, without the extensive aid of computers – are good at their work. Their structural assignments are usually correct; however, depending on the complexity of the structure of the unknown, conventional structure elucidation based largely on spectroscopic data can be a time-consuming process even in the hands of an expert in the field. When their assignments are incorrect, more often than not, the correct structure is one equally compatible with all of the available evidence, but overlooked by the chemist.

As the demand for structure elucidation increased, in particular in the pharmaceutical industry, it became evident that the time required by the chemist to fully interpret the collective spectroscopic data and manually reduce the inferred structural information to one or more plausible structures was a limiting factor in productivity. Thus, as new generations of spectrometers were delivering high quality spectroscopic data at even faster rates, productivity failed to increase. Since structure elucidation involves the analysis and processing of large amounts of data – spectroscopic data in this case – the computer, quite naturally, has been the centerpiece of numerous efforts to augment the productivity of the chemist and spectroscopist.

The development of computer-based tools for structure elucidation, which began almost 40 years ago, has not been easy because the goal is an especially ambitious one, namely, the creation of a computer program capable of simulating a high degree of human intelligence. Computer programs of varying sophistication have been described which are capable of executing each of the three major stages of structure elucidation. In some cases, the three stages (cf. Section 1.5) have been seamlessly incorporated into a single program.

In this volume, the reader is introduced to the structure generator Assemble 2.1. This program generates from the molecular formula all constitutional isomers that are compatible with the structural information. Its usage can enhance the structure elucidation and, perhaps most importantly, it can serve as a quality control of the structure elucidation process in the sense that no valid structure is overlooked.

1.7   Abbreviations and Symbols

CI	Chemical Ionization
comb	combination vibrations
COSY	Correlation SpectroscopY
d	doublet
dd	doublet of doublet
ddd	doublet of doublet of doublet
ddt	doublet of doublet of triplet
DEPT	Distortionless Enhancement by Polarization Transfer
dm	doublet of multiplets
DMSO	dimethyl sulfoxide
dq	doublet of quartet
dt	doublet of triplet
EI	Electron Impact ionization
EXSY	EXchange SpectroscopY
HMBC	Heteronuclear Multiple Bond Correlation
HSQC	Heteronuclear Single Quantum Correlation
NOE	Nuclear Overhauser Effect
NOESY	Nuclear Overhauser Enhancement and exchange SpectroscopY
q	quartet
s	singlet
st	stretching vibration
t	triplet
td	triplet of doublet
tt	triplet of triplet
u	mass unit (Dalton)
$w_{1/2}$	signal width at half height

2 Computer-Based Structure Generation

2.1 Introduction

As indicated earlier, of the three major stages of structure elucidation (Chapter 1.5), structure generation is not only the most tedious and least appealing for the chemist to carry out, but it is also prone to error and omission. Thus, it is not surprising that some of the earliest work in the area of computer-assisted structure elucidation was devoted to the development of structure generators.

Structure-generating procedures of practical value should meet three requirements. First, the procedure should be exhaustive. There are numerous examples of compounds with incorrect structural assignments in the primary literature, not because the structure is inconsistent with the evidence, but because another equally compatible structure was overlooked. Computers, unlike chemists at times, have no preconceived ideas about plausible structure types. Also, unlike chemists, computers, if properly programmed, excel in accuracy in the performance of repetitive, tedious tasks. Second, the program should execute efficiently, that is, on a time scale that enhances, not detracts from user productivity. Third, the output should be a set of irredundant structures.

Not unexpectedly, many structure-generating programs were designed to mimic an approach taken by the chemist, namely, connecting the substructural fragments predicted to be present in all possible ways. Such structure generators are classified as *Structure Assemblers*. In their basic operation, they can be described as procedures to systematically search for all valid connections between the *residual bonding sites* of the required substructures. If together, these required substructures do not account for all of the atoms in the molecular formula, which the program (like the chemist) requires; individual, unaccounted-for atoms ("naked" atoms) are treated as "substructural fragments" by the program. The set of substructures – single atoms included – represents a *partial structure* of the unknown. Structure generation can then be viewed as the process of *expanding* a partial structure into all complete molecular structures compatible with it. Depending on the richness of the information content of the partial structure (naked atoms, for example, are not very rich in information), expansion may lead to a single structure, very many structures, or somewhere in between. If the number is greater than one, but not too large, an examination of the generated structures by the chemist can serve as an invaluable guide in designing the most efficient experimental strategy to narrow the choices to the correct structure. If the number of structures is large, additional structural information will be required to proceed.

A number of structure-assembly-based structure generators have been developed to the stage of general utility. The application of one of these programs, Assemble, is described here. This program, one of the earliest to be developed, has undergone a number of revisions over the years and the current version serves as a versatile, computer-based tool for structure elucidation. The output of Assemble is an exhaustive and irredundant set of molecular structures – constitutional isomers – each of which is compatible with the molecular formula and all of the structural information contained in the input to the program.

The utility of the program can be illustrated by an application to the solution of a real-world structure elucidation problem. The antibiotic actinobolin was isolated from a fermentation brew. The compound has the molecular formula $C_{15}H_{22}N_2O_7$. Based on spectroscopic data and chemical evidence, the presence of a set of substructures in the intact compound was inferred by the chemist (Fig. 2.1). This set of substructures can be thought of as a partial structure of the unknown. Assemble was used to expand that partial structure into all molecular structures compatible with it. However, the structure generation was constrained by the input of some additional information deduced by the chemist. The presence of a 1,3-dicarbonyl unit was required, but compounds containing an aldehyde, a carboxylic acid, or a peracid functional group were forbidden. In addition, compounds with more than two hydroxyl groups were also forbidden. Assemble generated six structures, all of which were ketolactones (Fig. 2.2). Note that although diketones were not excluded by the input, none were generated. The information that requires this exclusion is intrinsic in the input.

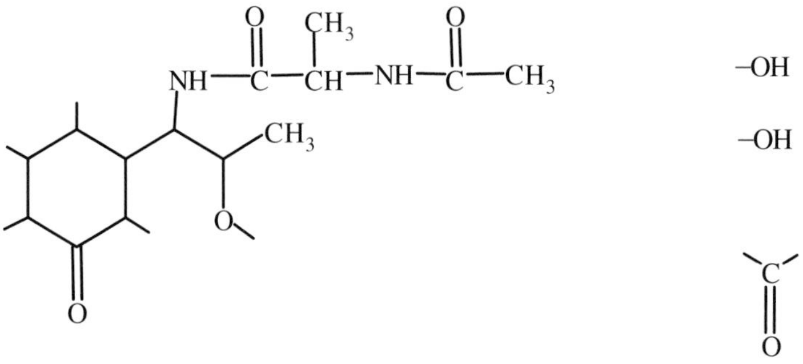

Fig. 2.1: Inferred substructures of actinobolin

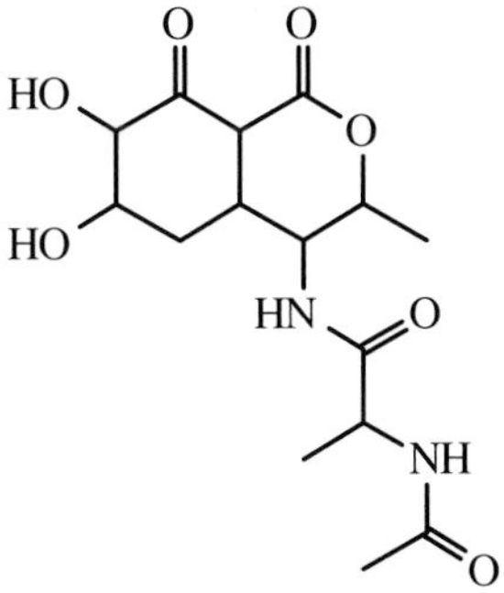

Fig. 2.2: Structures generated by Assemble 2.1 for $C_{15}H_{22}N_2O_7$ with the substructures shown in Fig. 2.1

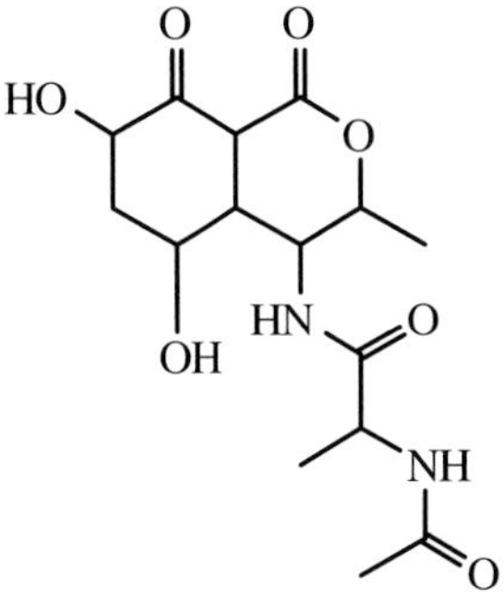

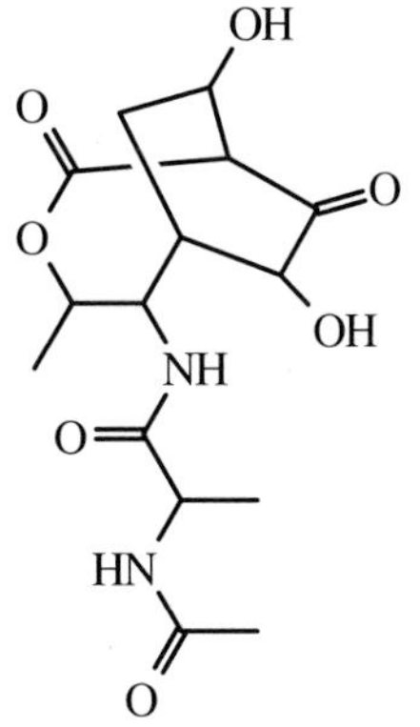

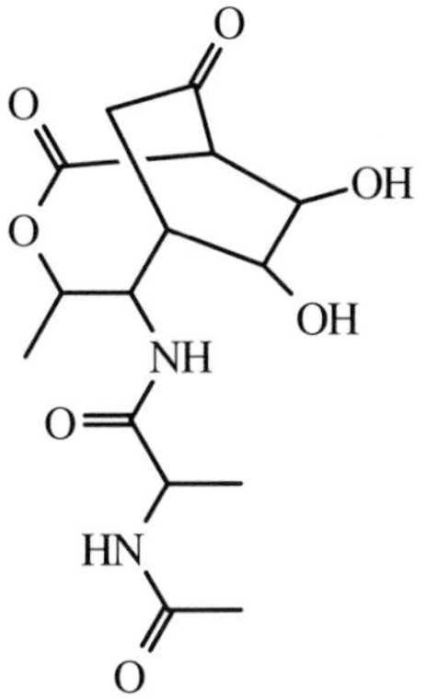

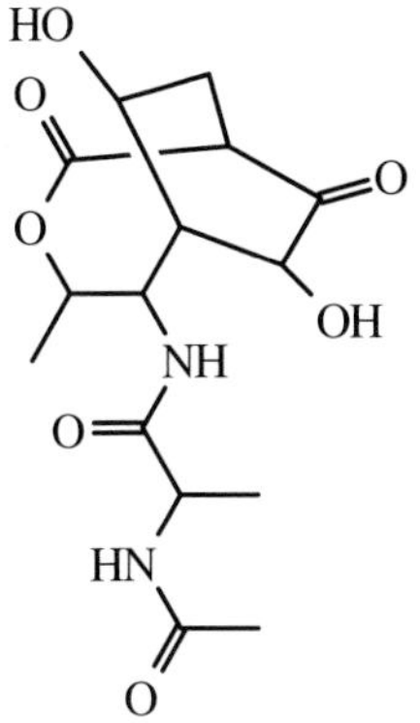

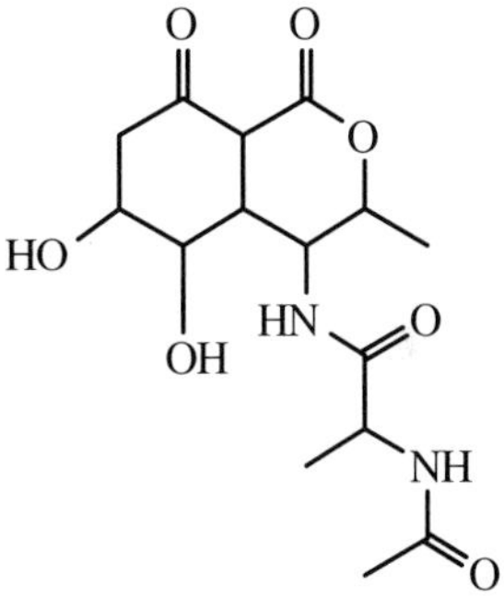

Since Assemble provides the assurance that no structure equally compatible with the input has been overlooked, the final assignment of structure was reduced to merely making a distinction between six and only six alternative structures. Examination of the six structures provided invaluable guidance in the design of a few simple experiments, which narrowed the choices and led to the correct assignment of the structure of actinobolin (structure **6**).

2.2 Principles and Procedures of Structure Assembly

The required information given to structure assemblers discussed here consists of three components: the *molecular formula* of the unknown, *nonoverlapping fragments* and *constraints*. Nonoverlapping fragments are substructures to be present in the unknown. Nonoverlapping means that an atom in any one fragment must not duplicate an atom in another fragment. Their use in the bond-forming process is straightforward. As the fragments are not allowed to overlap, all bonds in all such fragments can be formed at the beginning. This is not possible with *potentially* overlapping substructures. These have to be treated as constraints.

Constraints are *not* directly involved in the bond-making process. Instead, the information contained in the constraints applies restrictions on the bond-making process. Constraints therefore serve to reduce the size of the search space that must be explored and decrease the number of plausible structures generated.

A chemical compound may be viewed as *graph* in which the atoms are the vertices and the bonds are the edges which connect them. This is a useful concept since the principles of graph theory can then be applied to structure generation, that is, by expanding a partial structure into all complete molecular structures compatible with it (Chapter 2.1). The process of structure generation by means of structure assembly can be described using a pictorial representation called a Venn diagram (Fig. 2.3). The area within the rectangular border represents the *universal set of bonds*, that is, all bonds that are possible between the atoms in the molecular formula of the unknown. Thus, this diagram represents *all possible structural isomers* of a given molecular formula, each of which is described by a different subset of the universal set of bonds.

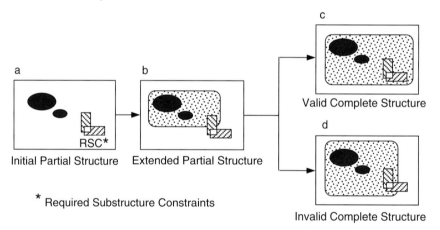

Fig. 2.3: Schematic representation of constrained structure assembly

In the example illustrated in Fig. 2.3, the initial problem state in structure assembly (the initial partial structure, Fig. 2.3a) consists of a set of *nonoverlapping* substructures (oval areas) and two overlapping substructures (rectangular areas) also required to be present in generated molecules. In this example, the latter two required substructures are *constraints*. Each substructure is represented as a subset of the universal set of bonds. The function of the structure generator is to search the solution space for all subsets of bonds corresponding to valid molecules, that is, molecules compatible with the nonoverlapping fragments and the imposed constraints. To begin the assembling process, a new bond is made between an available atom bonding site in one of the nonoverlapping fragments and a bonding site in the other fragment or some other atom in the molecule. This bond-making process continues in a stepwise fashion. However, at *each step*, the *expanded* partial structure (Fig. 2.3b) is tested for compatibility with each of the constraints. If a substructure is present, the constraint is met and no further testing is needed. Structure assembly proceeds until a valid molecular structure is generated (Fig. 2.3c). The same process then is repeated until *all* valid molecular structures have been generated.

In a given pathway of structure assembly, it is possible that a required substructure entered as a constraint may not be constructed until the very last bond completing a molecule is made. Or, the substructure may not be constructed at all, leading to an invalid complete structure (Fig. 2.3d). Thus, in structure assembly, the absence of a required substructure entered as a constraint cannot be established until the last bond completing a molecule is made; that is, a structure that fails to meet a particular required substructure constraint can only be excluded *retrospectively*. For this reason, as much information as possible should be entered as *nonoverlapping fragments* since that information is used *prospectively*.

Operationally, Assemble uses a depth-first search to expand the initial partial structure to an exhaustive set of compatible molecular structures. The simple "tree" shown in Fig. 2.4 illustrates a trace of the program execution. The root node (represented by a square) is the initial partial structure. Each triangular terminal node is a valid molecular structure. All other nodes are descendent nodes (circles), each of which is an expanded partial structure. Each edge of the tree represents the pathway traced by making a one-bond connection between available bonding sites of two atoms. Thus, at each level in the tree, each edge represents a different pathway of bond-making.

The integer labels on the edges of the tree describe the depth-first order in which the tree is traversed during structure assembly. A critical step in this process is node evaluation. Without "intelligent" node evaluation, Assemble would all too often follow branches of the tree leading either to invalid molecular structures or to duplicate structures, thereby using computational time unproductively. With such an inefficient process, the solution of real-world problems could require prohibitive amounts of computer time.

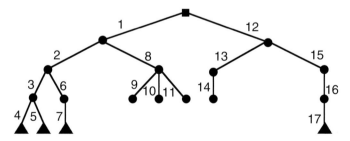

Fig. 2.4: A depth-first search

Node evaluation is based on two procedures. First, each expanded partial structure (node) is tested for compatibility with each of the imposed local and global constraints (cf. Chapter 2.3). Incompatibility with any one of the constraints leads to a termination of that pathway of bond-making, backtracking to the previously visited node in the tree, and exploration of a different pathway. Thus, not all branches in the tree lead to a complete molecular structure. For greatest computational efficiency, information should be used to terminate such unproductive branches at the highest possible level in the tree.

Second, at each node, symmetry relationships based on connectivity (*topological symmetry*) are perceived in order to avoid pathways leading to the generation of duplicate structures. Consider a simple example of the application of topological symmetry, an unknown run with an input consisting solely of the molecular formula, C_6H_{10}. At some point in traversing the tree, the six-membered cyclic partial structure (Figure 2.5a) will be encountered at a descendent node. Note that at this node, all carbon atoms in the unknown, but none of the hydrogens, have been incorporated in the partial structure. (Hydrogen atoms are fixed last in Assemble.) Therefore, each carbon atom has two remaining available bonding sites. One additional bond between two carbon atoms is required before the ten hydrogens are added to the remaining bonding sites to complete a molecular structure. Although only three discrete outcomes for the final carbon-carbon bond are possible (2.5b, c, d), many more pathways leading to those outcomes are possible, all but three of which are unproductive. Before bond formation is initiated, the topological symmetry algorithm of Assemble establishes two points. First, all of the six carbon atoms are topologically equivalent, i.e., they all have the same connectivity. Therefore, only *one* of the six sites is permitted to initiate bond formation. Second, the algorithm discerns that by selecting one of the sites to initiate bonding (for example, atom 1), that site is topologically differentiated from the other five sites. In the light of this information, the algorithm reexamines the topological symmetry of the five

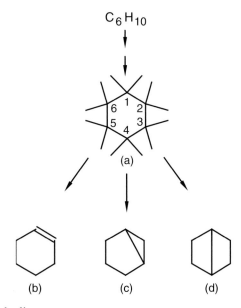

(a)

(b)   (c)   (d)

Fig. 2.5: Eliminating duplicate structures

terminating sites and determines that atom site 2 is equivalent to site 6, that site 3 is equivalent to site 5, and that site 4 is unique. As a result of this analysis, only three bond-making pathways are permitted and a large number of duplicate structures are never generated. Thus, Assemble "looks ahead" before making bonds and eliminates duplicate structures prospectively.

2.3 The Input to Assemble

The input to Assemble consists of three components: the *molecular formula* of the unknown, *nonoverlapping fragments*, and *constraints*. Elemental analysis combined with the spectroscopic data – in particular, the mass spectrum – usually leads either directly to the correct molecular formula or to a limited number of alternatives.

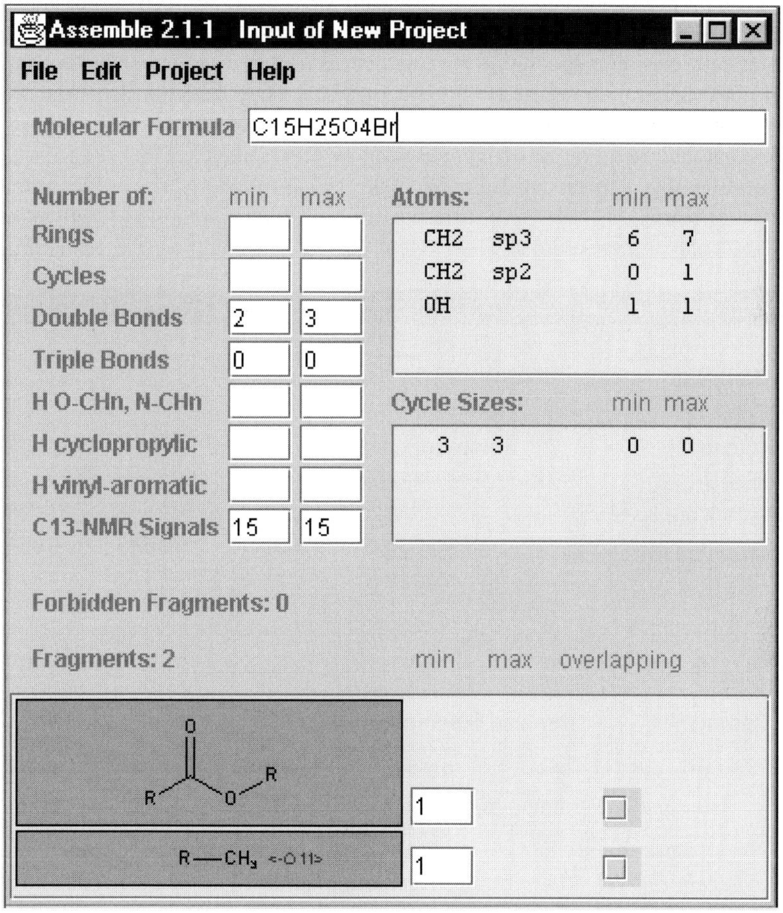

Fig. 2.6: Main input window of Assemble 2.1

Substructural information derived from an interpretation of the spectroscopic data should be expressed as nonoverlapping fragments whenever possible. This non-overlap restriction may at times present a problem to the user of Assemble in preparing the input. Since each fragment inferred is usually the result of an independent interpretation of selected spectroscopic data, it is entirely possible for two or more predicted substructures to contain atoms that are the same in the structure of the unknown. For example, consider a case where both a methoxy group ($O–CH_3$) and an ester function ($O=C–O$) are independently predicted to be present. Without evidence to the contrary, the oxygen atom within the methoxy group may or may not be the same as the corresponding oxygen atom of the ester function. Therefore, it would be inadvisable to enter those two substructures directly as nonoverlapping fragments. One of the fragments could be declared as potentially overlapping. However, there is a more efficient way to handle the situation by using *atom tags*, as described below.

Assemble accepts a broad range of constraints. These constraints are of two types, *local* and *global*. Local constraints provide additional information about the immediate environment of a specific atom in a substructure *entered as a nonoverlapping fragment*. Atom tag information is ignored if the fragment is declared as potentially overlapping. Global constraints, in contrast, are not site-specific. They describe structural features that characterize the compound as a whole. Most global constraints are entered by filling a number into the boxes in the main input window of Assemble (cf. Fig. 2.6).

Operationally, a local constraint is expressed by *tagging* a specific atom in a fragment. Assemble includes five different atom tags. Any atom in a fragment can be tagged with one or more atom tags.

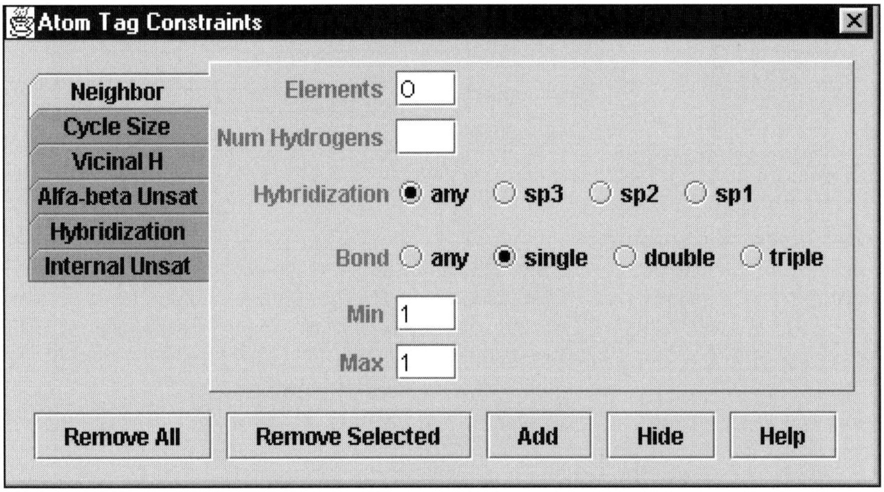

Fig. 2.7: Atom tag constraints window of Assemble 2.1. The neighboring atom tag is selected

The *neighboring atom tag* describes an atom that must join to a particular atom in a fragment (Fig. 2.7). The element type is required, but some additional information about the neighboring atom may be included: the hybridization of that atom, the number of attached hydrogens, the bond type by which it is joined to the tagged atom, and the minimum and maximum number of allowed neighboring atoms. The neighboring atom is *not* considered a part of the fragment. Therefore, that atom *may* overlap an atom in another required substructure declared as nonoverlapping. For preparing the input, the neighboring atom tag can serve as a means of including an atom in a fragment that potentially duplicates an atom in another fragment. In the example of the two potentially overlapping fragments described above (O–CH$_3$ and O=C–O), if the oxygen atom of one of the two fragments is expressed as an atom tag rather than a part of the fragment, both substructures can be entered as nonoverlapping fragments.

The *cycle tag* is used to designate the presence or absence of an atom in a fragment in a cycle of specified size. For example, if an infrared spectrum reveals the presence of a carbonyl group in the unknown whose bond angle is unstrained, the presence of the carbonyl carbon in a three- or four-membered cycle can be excluded using the cycle tag.

The *vicinal hydrogen tag* can be used to specify the total number of hydrogen atoms to be allowed on atoms contiguous to an atom in a fragment. For example, this tag is useful in cases where the number of hydrogens on carbon atoms adjacent to the carbon atom of a ketone carbonyl group has been determined by deuterium exchange experiments, which are monitored by ^{1}H NMR.

The *unsaturation tag* is used to require the presence or absence of an unsaturated linkage – for example a carbon-carbon double bond – joined to an atom in a fragment. Thus, a carbonyl group (C=O) entered as a fragment can be required or forbidden to be conjugated to some undesignated multiple bond linkage.

At times it may be desirable to designate the hybridization of an atom in a fragment which possesses residual valence allowing more than one state, for example, the terminal carbon atom in the fragment CH$_3$C. The *hybridization tag* serves this purpose.

Assemble includes seven global constraints several of which were designed to take advantage of structural inferences that can be derived from spectroscopic data. One of the constraints is a fragment declared as potentially overlapping. The number of occurrences can be specified as a number or a range. In particular, the substructure can be forbidden by specifying both minimum and maximum as zero.

The three *unsaturation constraints* – "Rings", "Double bonds constraint", and "Triple bonds" – are used to control the number of double bond equivalents to be expressed as rings, double and triple bonds, respectively. The *cycle constraints* are "Cycles" and "Cycle Sizes". "Cycles" controls the number of cycles of any size to be permitted, "Cycle Sizes", the number of cycles of specified size. In contrast to a ring, a cycle does not necessarily correspond to a double bond equivalent. A cycle is defined as a sequence of atoms A1–A2–...–An, where the atoms A1 and An are identical. The two rings of naphthalene account for 2 of its 7 double bond equivalents. However, naphthalene has three cycles, two of which are six-membered and one is ten-membered.

The "Atom" constraint serves to enter the number of atoms of an element in the unknown with a particular hydrogen multiplicity and, optionally, with a specified coordination number. For example, ^{13}C NMR spectroscopy is a source of such information for carbon atoms.

The constraints "H O–CH$_n$, N–CH$_n$", "H cyclopropylic" and "H vinyl-aromatic" control the number of hydrogens of defined type. Assemble recognizes cyclopropyl hydrogens, vinyl/aromatic hydrogens and hydrogens on carbon atoms bearing electronegative atoms such as oxygen, nitrogen, and halogen.

The constraint "C-13 NMR Signals" predicts the *number of signals* expected in the ^{13}C NMR spectrum of each molecular structure generated. A structure is considered to be valid if the predicted number of signals matches the observed number within the range defined by the user.

Assemble knows a number of strained features which give rise to molecular instability. The detection of some or all of these features can be enabled. Assemble terminates molecule assembly if strained features are detected. Thus, highly strained structures never appear in the output.

The breadth of constraints available in Assemble often allows the available structural information to be entered in more than one way. Thus, input preparation requires some thought on the part of the user. This is especially true in entering required substructures as *nonoverlapping fragments*. Initially, the user attempts to discern potential sites of overlap between the fragments, a task that may not be trivial. In cases where *one* atom in a fragment could possibly duplicate an atom in another fragment, one or the other atom can be expressed as a neighboring atom tag. Since the contents of an atom tag are not considered to be a part of a fragment, the desired information can be included without violating the non-overlap rule. With this application of the neighboring atom tag, there is no information loss.

In cases where potential overlaps can not be readily discerned or treated using the neighboring atom tag, one or more of the required substructures must be entered as a constraint since constraints are not subject to the non-overlap requirement. As a general rule, the more information-rich, potentially overlapping substructures should be entered as fragments. Less information-rich, potentially overlapping substructures are entered as constraints. This approach is based on the fact that a required substructure entered as a *fragment* is used more efficiently in program execution than when entered as a *constraint*. For the same reason, in preparing the input to Assemble, as much information as possible should be entered as *fragments*. Atom tags may be freely used wherever applicable to add information content since atom tags, although constraints, are part of fragments and used efficiently in structure generation.

2.4 Editing the Output of Assemble

The result of Assemble can be postprocessed much in the same way as an initial problem. All the *global constraints* are available to further constrain the structural variety. This allows the user to provide only the most trivial and certain information at the beginning, possibly creating a large number of structures. In the following steps, various assumptions of lesser certainty can be entered working on the set of initially created structures. As the initial information need not be given again, input creation is very fast. In addition, the computational effort is minimal. Postprocessing may be repeated as many times as required.

3   Problems

3.1 Problem 1

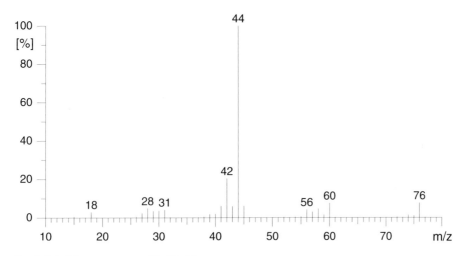

Fig. 3.1.1: Mass spectrum, EI, 70 eV

Fig. 3.1.2: IR spectrum, solvent CHCl₃, cell thickness 0.2 mm

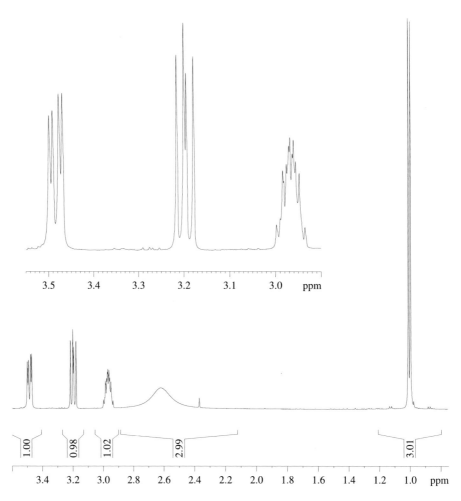

Fig. 3.1.3: ^{1}H NMR spectrum, 500 MHz, solvent CDCl$_3$

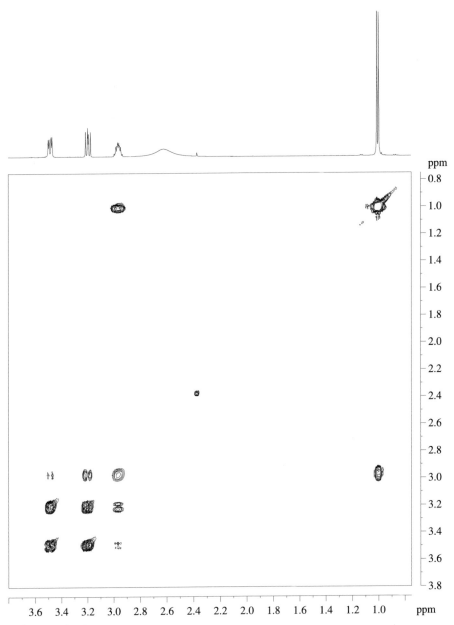

Fig. 3.1.4: ^{1}H,^{1}H COSY spectrum, 500 MHz, solvent CDCl$_3$

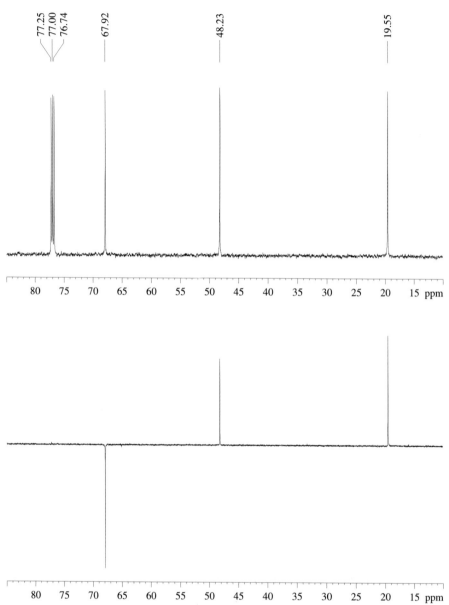

Fig. 3.1.5: ^{13}C NMR spectra, 125 MHz, solvent CDCl$_3$. Top: proton-decoupled; bottom: DEPT135

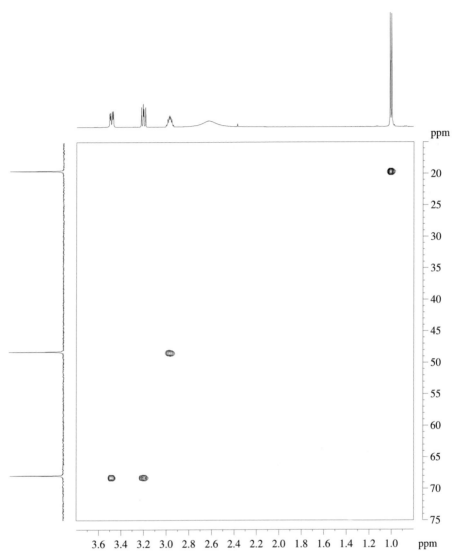

Fig. 3.1.6: $^{13}C,^1H$ HSQC spectrum, solvent $CDCl_3$

3.1.1 Elemental Composition and Structural Features

The relatively sharp band in the IR spectrum (Fig. 3.1.2) at 3630 cm^{-1}, together with the broad bands extending between 3600 and 2400 cm^{-1} indicate a partially hydrogen bonded hydroxyl group. The former band is assigned to the free hydroxyl group, whereas the latter corresponds to the associated form. An amino or imino group, if present, would be almost completely masked by the hydroxyl absorption. However, the weak maximum at 3370 cm^{-1} might be taken as indication of an NH (or NH$_2$) group being due to the stretching vibration of the (nonassociated) NH. No obvious assignment is possible for the band at 1590 cm^{-1}. Because of its shape, skeletal vibrations from an aromatic ring can be excluded. On the other hand, its intensity is too low for a carbonyl band. If NH is indeed present, it might be its deformation vibration.

In the mass spectrum (Fig. 3.1.1), the peak corresponding to the highest mass occurs at m/z 76. This cannot be the molecular ion, since the mass difference to the next important peak at m/z 60 is 16, which would correspond to methane, oxygen, or NH_2. Formation of such fragments from the molecular ion is, however, not very likely. We further note that the mass spectrum is dominated by ions at m/z 44 and 42.

The integration of the ^{1}H NMR peaks (Fig. 3.1.3) leads to the intensity ratios of 1:1:1:3:3 (from left). The doublet at δ 1.00 of intensity 3H indicates a methyl group connected to a CH, whose signal is located around δ 2.97. From the strong splitting of this multiplet, it is obvious that the methine proton is further coupled with other protons, resonating at δ 3.48 and 3.20.

The proton connectivities can be easily derived from the COSY (COrrelation SpectroscopY) map (Fig. 3.1.4). The methyl protons are coupled to the CH at δ 2.97, whereas this proton couples with the other two protons, which in turn couple with each other.

The broad signal at δ 2.62 of intensity 3H indicates exchangeable protons, e.g., of OH and/or NH. The presence of an OH group was inferred from the IR spectrum (see above).

The three signals in the ^{13}C NMR spectrum (Fig. 3.1.5, top) at δ 19.6, 48.2, and 67.9 can be assigned to a methyl, methine, and methylene group, respectively (cf. Comments), on the basis of the DEPT135 spectrum (Fig. 3.1.5, bottom). The chemical shift values for CH and CH_2 indicate that both groups are substituted by a hetero atom.

3.1.2   Structure Assembly

The elements found so far sum up to a partial molecular formula of C_3H_9O corresponding to 61 u. The presence of an odd number of hydrogen atoms demands an odd nominal molecular mass, if no other elements of odd atomic mass occur. At the same time, it indicates that an odd number of nitrogen atoms must be present. The elemental composition, therefore, is C_3H_9ON, with $M_r = 75$, for which Assemble 2.1 finds a total of 21 possible isomers.

Check with Assemble 2.1
Start Assemble 2.1, enter the molecular formula as the only piece of information, and generate all possible isomers. Consult the Tutorial (Chapter 5) if you should encounter difficulties with the program. Then, enter successively the Atom Constraints demanding the presence of 1 sp^3 CH_3, 1 sp^3 CH_2, and 1 sp^3 CH group. By simultaneously applying all three constraints, you will find only three possible isomers.

From the NMR spectra, the following partial structure can be derived:

$$
\left.
\begin{array}{c}
CH_3 \\
| \\
CH-Y \\
| \\
CH_2-X
\end{array}
\right\} \ 3H
$$

One of the two hetero atoms has to be an oxygen, as the IR spectrum requires a hydroxyl group. It must be connected to the methylene group according to the chemical shift of δ 67.7 in the ^{13}C NMR spectrum (see Comments). The nitrogen atom accomodates the other two exchangeable protons as a primary amino group. Thus, the constitution of the unknown compound is:

$$
\begin{array}{c}
CH_3 \\
| \\
CH-NH_2 \\
| \\
CH_2-OH
\end{array}
$$

The molecular mass is 75 u. The last peak at m/z 76 in the mass spectrum is, therefore, due to the protonated molecular ion. Such protonated molecular ions are commonly observed with primary aliphatic amines. The unassigned band at 1590 cm^{-1} in the IR spectrum can now also be explained as being due to the deformation vibration of the primary amino group.

3.1.3 Comments

3.1.3.1 Mass Spectrum

The fact that even-mass fragments dominate the mass spectrum and the significant peak at m/z 30 could be taken as primary diagnostic evidence for the presence of nitrogen. A very intense fragment of even mass within the nitrogen series m/z 30, 44, 58, ... is always suggestive of a saturated aliphatic amine residue. Analogously, the significant peak at m/z 31 is a reliable indicator of singly bonded oxygen.

If a mass spectrum (as the present one) ends with a cluster of weak signals, the assignment of a specific molecular mass becomes more ambiguous because the evaluation of isotope peak intensities is difficult or impossible. Consideration of the mass differences between all observable fragments becomes more critical and a trial and error analysis is usually performed in order to find the most convincing solution. Since, apart from hydrogen, 15 (CH_3) is the smallest chemically reasonable mass difference, and, in addition, loss of CH_3 radical is by far the most common primary fragmentation, the first attempt in such an analysis is to assume a molecular mass that is by 15 units higher than the largest significant fragment mass, in the present case 60 + 15 = 75. Under this assumption, the rationalization of the signal cluster around m/z 75 becomes straightforward because m/z 76 is to be interpreted as protonated molecular ion and m/z 74 as product of a deprotonation reaction, while m/z 58, 57, and 56 correspond to water elimination products.

Protonation of the molecular ion is a highly probable process anyway if aliphatic nitrogen and hydroxyl functions are indicated by other evidence. The tendency to protonate is especially pronounced in aliphatic amines, nitriles, and esters, though other polar groups are subject to such reactions and also often exhibit intensities of first isotope peaks of their molecular ions that are higher than required by elemental composition and natural isotope distribution. The assumption that m/z 76 represents the molecular ion provides less satisfactory explanations for the spectral features. A mass difference of 16 units to the first significant fragment is chemically not entirely unreasonable, but restricted to a few specific types of structures. In addition, loss of one and two hydrogen atoms with equal probability would be a quite unusual feature. The signal at m/z 18 (ionized water) should not be used as a structural argument, regardless of its intensity, because water is always present adsorbed on the sample or instrument surfaces and cannot be distinguished from water formed by degradation of the compound under study.

3.1.3.2 Infrared Spectrum

For simple molecules, as the one considered here, it is often possible to assign most major IR absorption bands by using correlation tables and reference spectra.

The OH stretching vibration of the free hydroxyl group appears at 3630 cm^{-1}. The free NH$_2$ group exhibits two stretching vibrations, one of the asymmetric mode at 3370 cm^{-1}, the other of the symmetric mode expected at 3300 cm^{-1}. There are, indeed, very small peaks discernible at these frequencies. The vibrations of the associated OH as well as of the NH$_2$ give rise to the broad band between 3600 and 2400 cm^{-1}. It is uncommon for simple alcohols and amines to have the stretching vibrations of the associated species extending much below 3000 cm^{-1} in dilute solutions. In the present case, this is most probably due to the formation of intramolecular as opposed to intermolecular hydrogen bonds. Another possibility is that the amino group is protonated to some extent. The anion may be chloride from the decomposition of chloroform or carbonate formed by reaction with carbon dioxide and moisture from the ambient air. The group of bands between 3000 and 2800 cm^{-1} is, of course, due to the various CH stretching modes. At 1590 cm^{-1}, we have the NH$_2$ deformation vibration, as already stated in the foregoing. At 1460 cm^{-1}, we find CH$_2$ deformation and CH$_3$ asymmetric deformation vibrations. The CH$_3$ symmetric deformation absorbs at 1380 cm^{-1}, and the low intensity band at 1350 cm^{-1} is ascribed to the deformation of the CH group bearing the nitrogen atom. From 1270 to 1200 cm^{-1}, the spectrum is masked by solvent absorption and can, thus, not be interpreted. Skeletal vibrations including CN and CO stretching modes are expected in this region. The strong absorption at 1040 cm^{-1} is most probably due to CO stretching. Below 1000 cm^{-1}, we have various bands of lesser intensity ascribed to OH deformation.

In those spectral regions where the solvent exhibits strong absorption bands, the detection system of the spectrometer receives little light. The apparent transmittance is, therefore, unpredictable and depends primarily on parameters not controlled by the operator. The recorder pen may drift to either side or remain stable. Moreover, it may also follow an erratic trace which, by chance, can have the appearance of a real absorption band. In most cases, however, regions of strong solvent absorption are readily identified by oddly shaped bands. Single beam Fourier transform instruments usually insert a horizontal line in these regions. This is well exemplified in the present spectrum for the bands between 1250 and 1180 cm^{-1} as well as between 815 and 660 cm^{-1}.

3.1.3.3 ^{1}H NMR Spectrum

The two amine and one hydroxy protons give rise to a broad line at δ 2.62. Usually, the exchange between amine and alcohol protons is fast on the NMR time scale and the protons are equivalent by a kinetic mechanism. Strong hydrogen bonding presumably contributes to the somewhat high value of the observed weighted average of the chemical shift values.

Since the molecule is chiral, the two methylene protons are diastereotopic and part of an almost first order A_3XYZ spin system. The vicinal coupling constants of the two diastereotopic protons with the methine proton are quite different indicating a preferred conformation, probably stabilised by intramolecular hydrogen bonds.

The difference of the heights of the two 4-line systems can be interpreted in terms of different line widths which can be explained to be a consequence of a coupling to the OH proton. Normally, CH–OH couplings are not visible in the ^{1}H NMR spectra because of fast intermolecular exchange of OH protons. Very strong hydrogen bonding, e.g., with dimethyl sulfoxide as solvent, reduces the exchange rate to such an extent that the couplings become visible.

In the present case, the intramolecular hydrogen bond is not strong enough to make the coupling obvious but is sufficient to cause line broadening. Vicinal couplings strongly depend on the dihedral angle, showing maxima at 0 and 180° and minima at 90°. Owing to the preferred conformation, there is a significant difference between the two HO–CH coupling constants and, therefore, the line widths of the two methylene proton signals.

The two tiny doublets near δ 1.13 and 0.87 are due to ^{13}C-^{1}H coupling (^{13}C satellites). Their distance from the chemical shift value of the corresponding protons in absence of ^{13}C isotopes equals half the direct ^{13}C-^{1}H coupling constant (the isotope effect of ^{13}C on the ^{1}H-chemical shift is negligible in routine analysis). Generally, each signal is accompanied by spinning side bands in addition to the ^{13}C-satellites. The spacing of these side bands is also symmetric around the main signal and equals the spinning frequency of the sample tube or an integral multiple thereof. They are caused by magnetic field inhomogeneities and, in general, exhibit an intensity of < 0.5–1% of the main band.

3.1.3.4 ^{13}C NMR Spectrum

In ideal cases, the interpretation of DEPT (Distortionless Enhancement by Polarization Transfer) spectra is very simple: CH$_3$ and CH carbon atoms are positive, and CH$_2$ has negative signals in the DEPT135 spectrum. No signals appear in DEPT spectra for carbons that are not directly attached to hydrogen atoms. However, multipulse techniques for identifying the number of protons directly attached to a carbon atom assume a single fixed value for all $^1J_{C,H}$ (one-bond) coupling constants. If the deviation is huge, the signal may be outright wrong, e.g., in the case of acetylenes.

Comparing the higher sensitivity of the ^{1}H-detected two-dimensional ^{13}C,^{1}H COSY (Heteronuclear Single Quantum Coherence, HSQC) method with that of the DEPT, the former measurement is preferred nowadays. Assignment of the corresponding ^{1}H and ^{13}C signals can be taken directly from the HSQC spectrum (Fig. 3.1.6). There is no cross peak with the signal at δ 2.62 in accordance with the assignment of this signal as XH protons, where X is not a C atom.

3.1.3.5 Presentation of NMR Data

^{1}H NMR (500 MHz, CDCl$_3$): δ = 3.48 (dd, J = 10.6, 3.9 Hz, 1H, OC$\underline{H}_a$H$_b$), 3.20 (dd, J = 10.6, 7.9 Hz, 1H, OCH$_a$$\underline{H}_b$), 2.97 (m, 1H, NCH), 2.62 (broad, 3H, OH, NH$_2$), 1.00 (d, 3H, CH$_3$). ^{13}C NMR (125 MHz, CDCl$_3$): δ = 67.9 (OCH$_2$), 48.2 (NCH), 19.6 (CH$_3$).

3.2   Problem 2

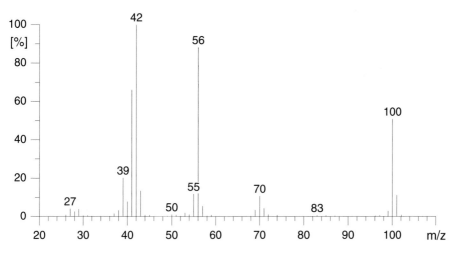

Fig. 3.2.1: Mass spectrum, EI, 70 eV

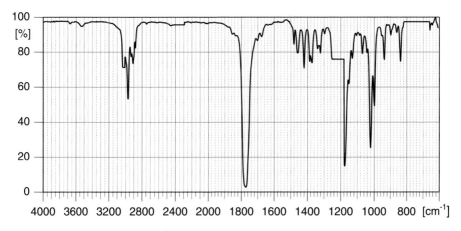

Fig. 3.2.2: IR spectrum, solvent CHCl₃, cell thickness 0.2 mm

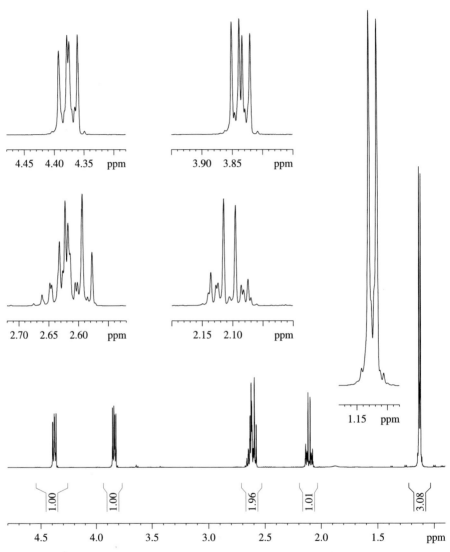

Fig. 3.2.3: ^{1}H NMR spectrum, 500 MHz, solvent CDCl$_3$

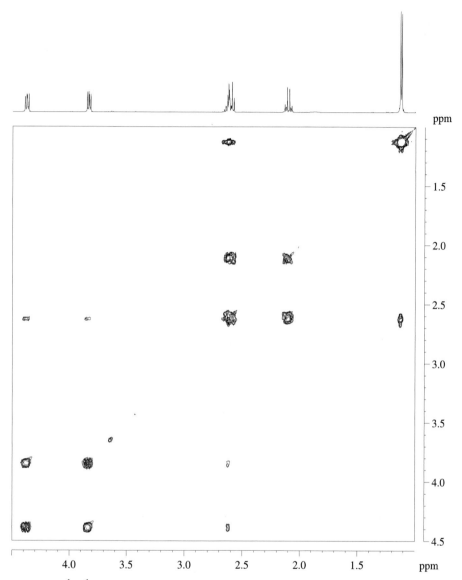

Fig. 3.2.4: ^{1}H,^{1}H COSY spectrum, 500 MHz, solvent CDCl$_3$

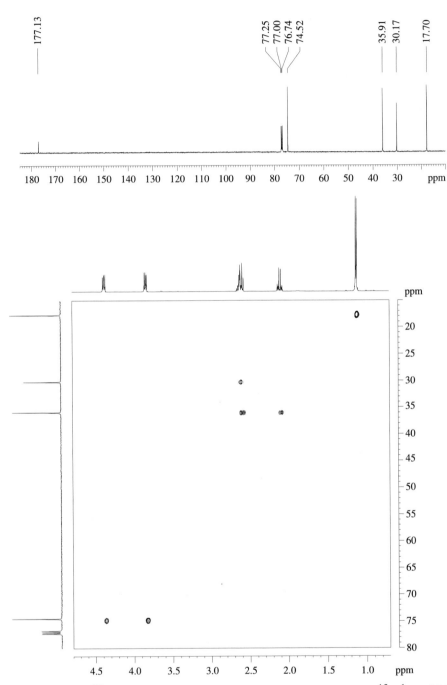

Fig. 3.2.5: Top: ^{13}C NMR spectrum, 125 MHz, solvent CDCl$_3$; bottom: ^{13}C,^{1}H HSQC spectrum

3.2.1 Elemental Composition and Structural Features

Integration in the ^{1}H NMR spectrum (Fig. 3.2.3) gives an intensity ratio of 1:1:2:1:3 (from left to right) and corresponds to a total of 8n protons (n = 1, 2, ...). The doublet at δ 1.12 of intensity 3H indicates a methyl group attached to CH.

In the ^{13}C NMR spectrum (Fig. 3.2.5, top), the signal at δ 177.1 indicates an O=C–X group, X being a hetero atom other than sulfur, otherwise the chemical shift would be considerably higher. The one-bond connectivities between the ^{1}H and ^{13}C atoms can be simply read from the cross peaks in the two-dimensional ^{13}C,^{1}H-correlated HSQC spectrum (Fig. 3.2.5, bottom), where the x axis represents the ^{1}H and the y axis the ^{13}C chemical shifts. On the basis of the presence of two cross peaks, the ^{13}C signals at δ 74.5 and 35.9 can be assigned to methylene groups. The signal at δ 30.2 belongs to a CH since it correlates with one single proton (intensity, 1H) at δ 2.63.

The presence of a carbonyl group is confirmed in the IR spectrum (Fig. 3.2.2) by the C=O stretching vibration band at 1770 cm^{-1}. The strong absorption band at 1170 cm^{-1} is very likely due to a C–O–C stretching vibration. The range of ca. 1260–1190 cm^{-1} is biased by solvent absorption.

The elemental composition of the structural elements found so far amounts to C_5H_8O (84 u). As all fragments correspond to chemically reasonable mass differences relative to the last significant signal at m/z 100 in the mass spectrum (Fig. 3.2.1), we assume a molecular mass of 100. The difference between this molecular mass and the mass of the structural fragments found so far amounts to 16 u and must be assigned to an oxygen atom since no additional protons are available. The molecular formula thereby becomes $C_5H_8O_2$ and shows two double bond equivalents. One of them is taken care of by the carbonyl group, the other one must be due to a ring since no other double bond is indicated.

3.2.2 Structure Assembly

On the basis of the molecular formula $C_5H_8O_2$, the structure generator, Assemble 2.1, generates 1168 isomers These include many unreasonable structures.

Check with Assemble 2.1

Start Assemble 2.1, enter the molecular formula as the only piece of information, and generate all possible isomers. Choose Forbidden Fragments from the Edit menu. A new window pops up showing 7 kinds of strained structures that can be excluded group-wise. Calculate the number of possible isomers by forbidding all strained structures. Several further improbable structures, such as enols or peroxides, remain because they may occur in valid molecules and should, therefore, not be excluded globally.

Of course, it is possible to exclude improbable solutions by drawing the respective substructure and demanding that it must not occur. Click into the Fragments window (the lowest part of the Assemble input window) with the left mouse button, select Draw Fragment, and draw an enol with three open valences (symbol R). After completing the drawing, again press the left mouse button in the drawing window and use the pop-up window to add the substructure to Assemble

Fragments. In the appearing menu check Overlapping and set Min and Max to 0 (Fig. 3.2.6).

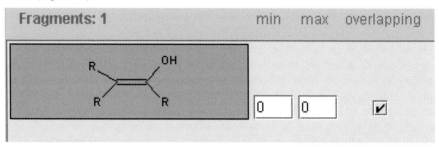

Fig. 3.2.6: Enol as a forbidden substructure in the input window of Assemble 2.1

With this input item, the number of possible isomers is reduced by 156. Of course, it is also possible to set both the min and max numbers of the entered fragment to 1 and generate the 156 enols. Similarly, other substructures such as peroxides can be excluded.

On the basis of all the pieces of information found under 3.2.1, Assemble 2.1 generates three structures, i.e., α-, β-, and γ-methyl-γ-butyrolactone. The third one can easily be excluded because of the ^{1}H and ^{13}C chemical shifts of δ 3.9–4.4 and 74.5, respectively, which indicate the presence of a CH_2O group. The two remaining structures are:

I   **II**

Structure **I** can be eliminated by evaluating the two-dimensional COSY NMR spectrum (Fig. 3.2.4), where both axes correspond to ^{1}H chemical shifts. Cross peaks in the spectrum (i.e., those that are not on the diagonal) mark the proton-proton coupling partners. The two methylene groups cannot be linked to each other because there are no cross peaks between them. It can be seen that the methyl signal correlates with the signal at δ 2.63 and this CH is coupled to all four remaining protons. Such an arrangement is present only in structure **II**, whereas in the case of **I**, the CH proton would correlate only with the methylene protons corresponding to the lower chemical shifts.

3.2.3   Comments

3.2.3.1 Mass Spectrum

The mass spectrum illustrates the fact that occasionally even-mass fragments can dominate in the absence of nitrogen if the compound contains to a large extent especially good leaving groups like CO and CO_2. The choice between **I** and **II** could safely be based on mass spectrometric evidence if accurate mass measurement showed that m/z 42 is mainly due to ketene radical cations ($CH_2C=O$ in **II**) rather than to C_3H_6 (in **I**).

3.2.3.2 Infrared Spectrum

The most conspicuous feature of the IR spectrum is the high frequency of the carbonyl stretching vibration. Such vibrations are shifted to higher frequencies primarily because of a decrease in the angle between the substituents and by substitution with electronegative groups. The world record in high carbonyl stretching frequencies is held by carbonyl fluoride, COF_2, which absorbs at 1928 cm^{-1}. In the compound at hand, it is the steric influence of the five-membered ring on the C–O–CO–C moiety that is responsible for the high frequency.

In IR spectra, the light absorption is commonly recorded as transmittance, expressed in %. Transmittance is the ratio between the intensity of the transmitted light in the sample beam and the intensity of the reference beam. According to Beer's law, transmittance ideally varies exponentially with the concentration c, the cell thickness l, and the extinction coefficient ε:

$$T = \frac{I}{I_0} = 10^{-\varepsilon c\, l} \tag{1.1}$$

Displaying in absorbance, $A = -\log T = \varepsilon c\, l$, becomes increasingly popular because it is directly proportional to concentration and path length.

3.2.3.3 ¹H NMR Spectrum

Due to the stereogenic center (methine), the protons of the CH_2–O group are diastereotopic and thus, at best, may accidentally have the same chemical shift (i.e., be isochronous). Here, the difference between the corresponding chemical shifts is ca. 0.6 ppm. This large difference is due to the vicinal methyl group. The influence of a methyl group on the chemical shift of protons in γ position strongly depends on the conformation. A synplanar conformation leads to a shielding effect of ca. 0.8 ppm, whereas an antiplanar one causes deshielding by ca. 0.3 ppm:

CH₃   H
$\delta = -0.8$ relative to   H   H

CH₃
                  $\delta = +0.3$ relative to   H
        H   H

In analogy, the CH_2CO protons can be assigned to the signals at approximately δ 2.60 and 2.11, while the methine proton has a chemical shift close to δ 2.63. Chemical shift arguments thereby allow to assign each one of the geminal methylene protons. No such assignment is possible here by considering vicinal coupling constants, very much in contrast to the situation encountered in case of six-membered rings (see Problems 6 and 15). In five-membered rings, because of their greater flexibility, no simple correlation exists between the relative magnitudes of the *cis* and *trans* coupling constants.

3.2.3.4 Presentation of NMR Data

^1H NMR (500 MHz, CDCl₃): $\delta = 4.38$ (dd, $J = 7.2, 8.9$ Hz, 1H, OCH$_a$H$_b$), 3.84 (dd, $J = 6.5, 8.9$ Hz, 1H, OCH$_a$H$_b$), 2.63 (m, 1H, CH), 2.60 (m, 1H, CH$_a$H$_b$), 2.11 (m, 1H, CH$_a$H$_b$), 1.12 (d, $J = 6.6$ Hz, 3H, CH₃). ^{13}C NMR (125 MHz, CDCl₃): $\delta = 177.1$ (C=O lactone), 74.5 (OCH₂), 35.9 (CH₂), 30.2 (CH), 17.7 (CH₃).

3.3 Problem 3

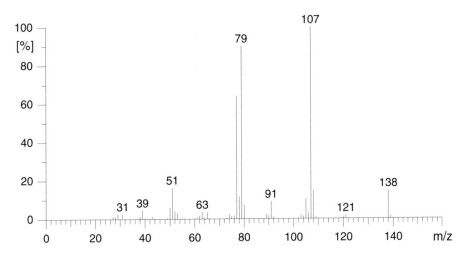

Fig. 3.3.1: Mass spectrum, EI, 70 eV

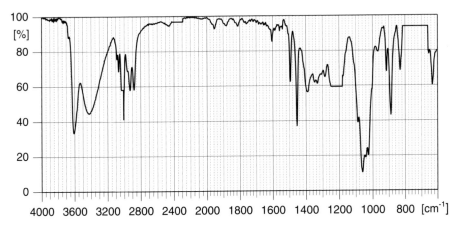

Fig. 3.3.2: IR spectrum, solvent CHCl₃, cell thickness 0.2 mm

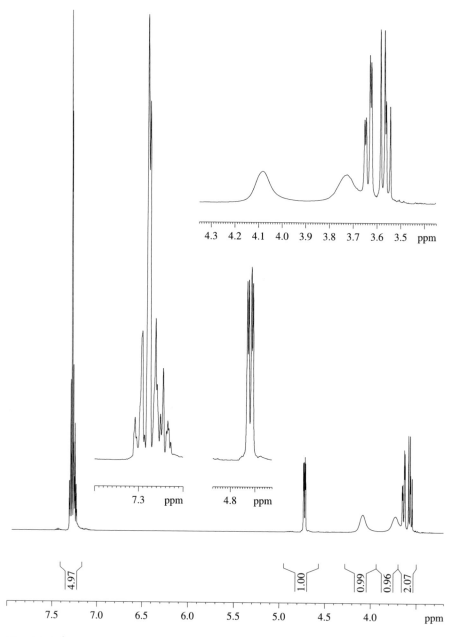

Fig. 3.3.3: ^{1}H NMR spectrum, 500 MHz, solvent CDCl$_3$

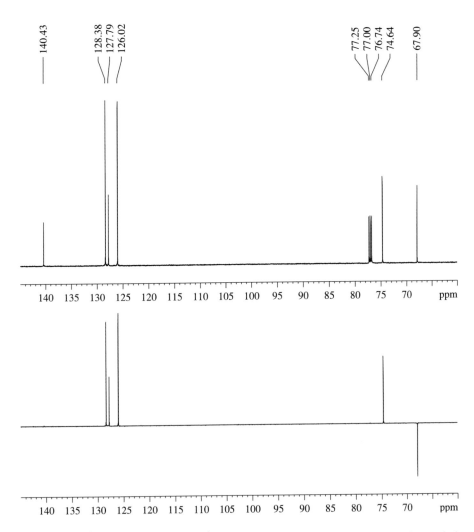

Fig. 3.3.4: ^{13}C NMR spectra, 125 MHz, solvent CDCl$_3$. Top: proton-decoupled; bottom: DEPT135

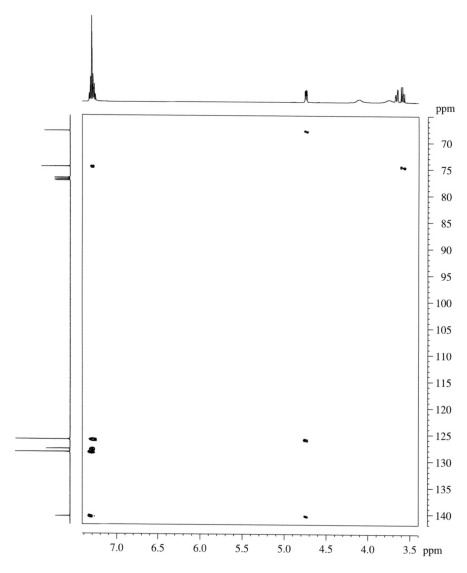

Fig. 3.3.5: $^{13}C,^{1}H$ HMBC spectrum, solvent $CDCl_3$

3.3.1   Elemental Composition and Structural Features

The IR spectrum (Fig. 3.3.2) indicates that one or more OH groups (3600 cm^{-1}, free OH st; 3425 cm^{-1}, associated OH st) and a phenyl ring, probably monosubstituted (2000–1600 cm^{-1}, comb; 1600 and 1500 cm^{-1}, skeletal vibrations), must be present and that there are no carbonyl groups in the molecule.

The mass spectrum (Fig. 3.3.1) ends with m/z 138. The differences to the other peaks being chemically reasonable, we tentatively assume a molecular mass of 138. The benzene ring is confirmed by the signals at m/z 39, 51, 63–65, 77. The significant peak at m/z 91 indicates that the phenyl group is probably attached to an sp^3-hybridized carbon atom.

The six signals in the ^{1}H NMR spectrum (Fig. 3.3.3) at δ 7.27 (m), 4.72 (dd, J = 8.6, 3.2 Hz), 4.08 (s, broad, $w_{1/2}$ = 37 Hz), 3.73 (s, broad, $w_{1/2}$ = 45 Hz), 3.64 (dd, J = 11.5, 3.2 Hz), and 3.56 (dd, J = 11.5, 8.6 Hz) correspond to a proton ratio of 5:1:1:1:2, i.e., a total of 10H. According to the chemical shift and intensity of the multiplet at δ 7.27, the phenyl group is monosubstituted.

The proton-decoupled ^{13}C NMR spectrum (Fig. 3.3.4, top) consists of six signals. As expected for a monosubstituted phenyl ring, four of them occur in the aromatic region between δ 126–140. The DEPT135 spectrum (Fig. 3.3.4, bottom) shows that three of them correspond to CH and one to a quaternary C. Two further signals appear in the aliphatic shift region, one for CH (δ 74.6) and the other for CH_2 (δ 67.9), both substituted with a strongly electronegative substituent. Since only 8 of the 10 protons are accounted for by the ^{13}C NMR spectrum, two hydrogen atoms must be attached to hetero atoms. Therefore, the presence of two OH groups can be assumed.

3.3.2 Structure Assembly

The results obtained so far show the presence of a monosubstituted phenyl ring, one methylene and one methine group, and two hydroxyls. This sums up to $C_8H_{10}O_2$ (138 u), indicating that all constitutional elements of the molecule have been identified. There is only one possibility of connecting them in a chemically meaningful way, namely:

Check with Assemble 2.1
Although this molecule is chiral, the aromatic carbon atoms in *ortho* and *meta* positions are isochronous because the phenyl group exhibits a fast rotation around the bond, which is an axis of symmetry for this group (see Chapter 4.3). Such a symmetry constraint is, in general, a powerful structural information. Based on the molecular formula $C_8H_{10}O_2$ alone, Assemble 2.1 generates 605 379 isomers. Only 19 389 of them have 6 signals in the ^{13}C NMR spectrum (this information can be entered in the input window). The number is further reduced to 8 566 if strained structures are excluded. Even so, some structures are uncommon and the calculation of atom coordinates is not always possible. You can copy such molecules into the structure editor JUME and unscramble them manually, if necessary.

The solutions can be ranked according to the ^{1}H and ^{13}C chemical shifts. Start Rank Output from the Project menu and select H NMR. Use Set Example to see the required input format. The program uses only simple first-order multiplicities.

In the ranked output, the correct structure appears as number 2 if the default setting is used, which sets structures for which no prediction is possible, in front of the others. However, more important than the ranking position is the number of structures for which the mean and maximal deviations between predicted and observed chemical shifts are so small that they could be valid (see Fig. 3.3.6). In the present case, only five of the 8 566 unstrained structures have mean deviations < 0.50 ppm and maximum deviations < 1.00 ppm.

Assemble 2.1 tries its best even if some of the chemical shifts cannot be estimated. The color of the text above the structure gives a hint about the quality of the spectrum estimation: *Red* means that at least one chemical shift could not be estimated because the information about the chemical environment was not available. The corresponding shift is, therefore, not used in the comparison with the experimental spectrum. Keep in mind that generated candidate structures may look weird. Very possibly this is the case with a red entry. *Magenta*: The estimation of at least one shift has not been possible since either the corresponding hydrogen atom is bonded to a hetero atom or the estimation, albeit of lower quality, is used for comparison with the experimental spectrum. *Blue*: All chemical shifts have been estimated satisfactorily.

Similarly, the ^{13}C chemical shifts can be used to rank the solutions (see Fig. 3.3.6). In the present case, the correct structure appears as the best one. Already the next one shows large mean (4.7 ppm) and maximum (22.1 ppm) deviations between estimated and measured shifts.

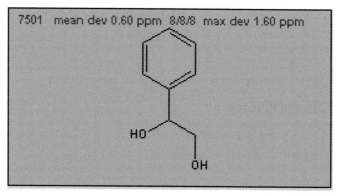

Fig. 3.3.6: Part of the ^{13}C NMR ranking output showing the structure number 7501 with its mean and maximum deviations between estimated and measured chemical shifts. The three numbers 8/8/8 signify, respectively, that 8 chemical shifts are estimated, 8 signals are predicted to appear in the ^{13}C NMR spectrum of the structure, and 8 experimental shifts are entered.

3.3.3 Comments

3.3.3.1 Mass Spectrum

As expected, the base peak formation results from benzylic cleavage, which is favored by the presence of the hydroxyl groups. The subsequent decarbonylation (m/z 107 →79) is typical of deprotonated benzyl alcohol cations. Water elimination from the molecular ion probably results in styrene oxide, which is known to lose CHO (Δm 29) after rearrangement to the isomeric aldehyde. This reaction sequence gives rise to m/z 91, as described in the following simplified formalism:

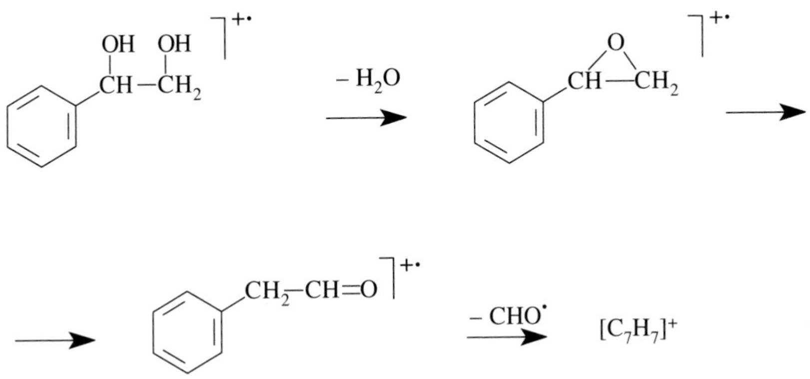

3.3.3.2 Infrared Spectrum

Prominent skeletal vibration frequencies for benzene rings are observed near 1600 cm^{-1}, between ca. 1500 and 1450 cm^{-1}, and near 700 cm^{-1}. The respective vibrations may be described as quadrant stretching, semicircle stretching, and sextant bending. The first two vibrations consist of two components each, which are often resolved in the spectrum.

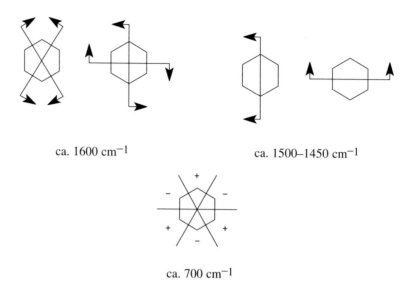

ca. 1600 cm^{-1}

ca. 1500–1450 cm^{-1}

ca. 700 cm^{-1}

The quadrant stretching vibrations near 1600 cm^{-1} are absent for centrosymmetric benzene compounds whose center of symmetry coincides with the center of the ring. In monosubstituted benzene rings, the intensity of the corresponding bands is high for substituents that are either strong electron acceptors or strong electron donors. For weakly interacting substituents, the intensity is low. In *p*-disubstituted compounds, the intensity is proportional to the difference between the electronic effects of the two substituents, whereas in *m*-disubstituted compounds it is proportional to their sum. The intensities in *o*-disubstituted derivatives are of intermediate size. The semicircle stretching bands near 1500 cm^{-1} are always present, although sometimes rather weak, e.g., if the substituents are carbonyl groups.

The band at 700 cm^{-1} is absent from IR spectra of benzene compounds exhibiting a center of symmetry in the ring center as well as of *o*-disubstituted compounds if the two substituents are identical. Even with non identical substituents, the intensities are low in *o*- and *p*- disubstituted compounds. High intensity absorptions are generally observed in mono-, *m*-di-, and symmetrically trisubstituted benzene compounds. The intensities of the respective bands observed in the IR spectrum of the present compound fit well into this scheme.

In benzene systems, a series of low intensity bands are observed between 2000 and 1660 cm^{-1}. These combination bands arise from various interactions between C–H deformation vibrations and skeletal ring vibrations. They show a characteristic pattern that depends on the number of vicinal hydrogen atoms on the benzene ring and, thus, are indicative of the substitution type. The position of the bands is rather variable but the overall pattern is reasonably constant. When assigning substitution types, it is, therefore, advisable to consult suitable reference spectra rather than numerical lists of band positions. In standard spectra, the intensity of these combination bands is generally too low to be of practical value. An exception are monosubstituted and *p*-disubstituted benzene rings. The former exhibit four equally spaced absorptions of distinct intensity, whereas the latter are characterized by one relatively strong and one much weaker absorption band. The correlation between the appearance of the spectrum in this range and the substituent pattern is quite reliable as long as there are no substituents that strongly interact with the π system of the ring. With such substituents (e.g., C=O, NO$_2$, C=C, etc.), the correlation becomes unreliable and may be misleading.

3.3.3.3 ^{1}H NMR Spectrum

As a consequence of the fast intermolecular exchange of the hydroxyl protons, their vicinal coupling with protons on the neighboring carbon atoms generally does not lead to observable splittings in the ^{1}H NMR spectrum. Traces of acids (generally present in deuterochloroform) or bases catalyze this exchange reaction. For very pure alcohols in acid- and base-free solvents, the exchange rate is often slow enough to make these couplings observable. In the present case, it is slow relative to the chemical shift difference between the two hydroxyl protons (0.35 ppm, i.e., 175 Hz) but fast relative to the vicinal coupling constants of ca. 7 Hz. Therefore, two distinct signals are observed, which are broadened due to the exchange, but no coupling to the vicinal CH$_2$ and CH is seen in the spectrum. Accordingly, the mean lifetime, τ [s], of the species is:

$$\frac{1}{175} < \tau < \frac{1}{7}$$

In CDCl$_3$, the two diastereotopic methylene protons have different chemical shifts (δ 3.64 and 3.56) and couple differently to the vicinal methine proton (J = ca. 8.6 and 3.2 Hz, respectively). This information could be the base of an educated guess about the prevailing conformation. The staggered conformations given below are considered:

For a torsion angle of 180° one expects a large coupling constant of the order of 10 Hz, whereas for 60° the coupling constant is only ca. 4 Hz. This rules out conformation **III** since it would demand two small vicinal coupling constants. A detailed calculation taking into account the influence of the substituents gives a slightly better fit for conformation **I**, which is also expected because of its possibility to form an intramolecular hydrogen bond.

3.3.3.4 ^{13}C NMR Spectrum

Generally, the line intensities in routine ^{13}C NMR spectra are not proportional to the number of carbon atoms represented by the respective signal (cf. Chapter 4.3). However, for carbon atoms directly attached to hydrogen(s), similar relaxation times and consequently similar line intensities are to be expected, even if the spectrum is recorded under conditions of partial saturation (provided that the C–H vectors exhibit comparable mobilities). Therefore, on the basis of its lower intensity, the signal at δ 127.8 can be assigned to the aromatic carbon atom in *para* position relative to the substituent.

The couplings, $^nJ(C,H)$ of 0–10 Hz for n = 2, 3 give rise to cross peaks in the $^{13}C,^1H$ HMBC (Heteronuclear Multiple Bond Correlation) spectrum (Fig. 3.3.5), which provides a complete and unambiguous signal assignment. Couplings over four bonds can only rarely be observed in HMBC spectra. The cross peak at 7.27/74.6 identifies the methine carbon through coupling over three bonds, whereas the methine proton (δ 4.72) gives correlations over two bonds to the methylene (δ 67.9) and to the quaternary C_{ipso} (δ 140.4) carbon atoms. A differentiation between the C_{ortho} and C_{meta} signals is feasible on the basis of the cross peak at 4.72/126.0.

3.3.3.5 Presentation of NMR Data (500 resp. 125 MHz, CDCl$_3$, δ)

Assignment	^1H (J)	^{13}C	HMBC responses (^{13}C partners)
CH$_2$	3.64, dd (11.5, 3.2 Hz),	67.9	CH
	3.56, dd (11.5, 8.6 Hz)		
CH	4.72, dd (8.6, 3.2 Hz)	74.6	CH$_2$, C$_{ipso}$, C$_{ortho}$
C$_{ipso}$	–	140.4	
CH$_{ortho}$	7.27, m	126.0	CH, (C$_{para}$)
CH$_{meta}$	7.27, m	128.4	C$_{ipso}$
CH$_{para}$	7.27, m	127.8	(C$_{ortho}$)
OH	4.08, s, broad, w$_{1/2}$ = 37 Hz		
	3.73, s, broad, w$_{1/2}$ = 45 Hz		

3.4 Problem 4

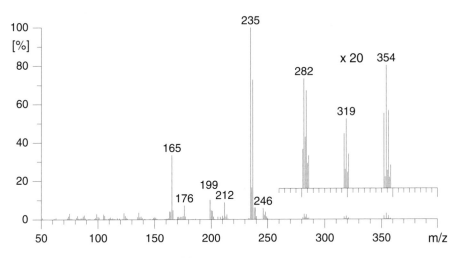

Fig. 3.4.1: Mass spectrum, EI, 70 eV

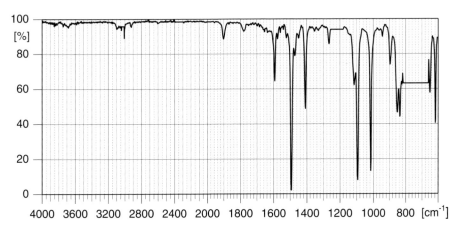

Fig. 3.4.2: IR spectrum, solvent CHCl₃, cell thickness 0.2 mm

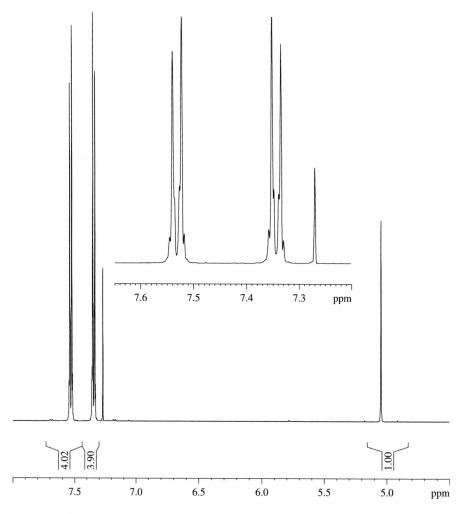

Fig. 3.4.3: ^{1}H NMR spectrum, 500 MHz, solvent CDCl$_3$

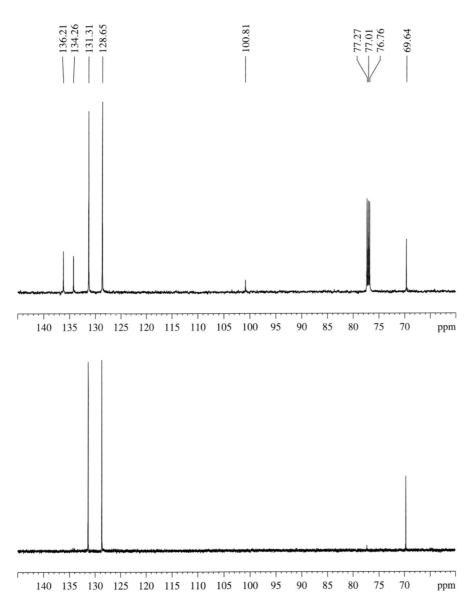

Fig. 3.4.4: ^{13}C NMR spectra, 125 MHz, solvent CDCl$_3$. Top: proton-decoupled; bottom: DEPT135

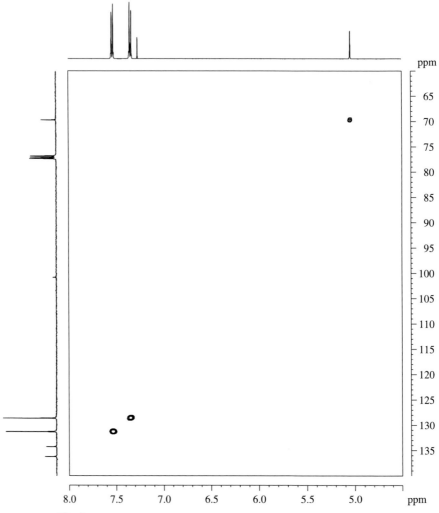

Fig. 3.4.5: ^{13}C,^{1}H HSQC spectrum, solvent CDCl$_3$

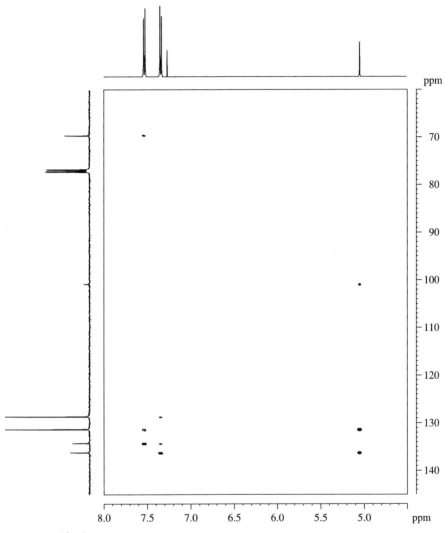

Fig. 3.4.6: $^{13}C,^1H$ HMBC spectrum, solvent CDCl$_3$

3.4.1   Elemental Composition and Structural Features

The high frequency range of the IR spectrum (Fig. 3.4.2) is practically empty. Weak combination bands at 1900 and 1780 cm^{-1} and sharp bands at 1600 and 1500 cm^{-1} indicate an aromatic system.

Three signals with an intensity ratio of 4:4:1 appear in the ^{1}H NMR spectrum (Fig. 3.4.3). The chemical shifts of δ 7.53 (dm, i.e., doublet of multiplets) and 7.34 (dm) indicate the presence of aromatic protons and the symmetry of the multiplets (spin coupling pattern *AA'XX'*), a symmetric ring substitution. Since the coupling pattern resembles an *AB* system and also because of the number of aromatic signals in the ^{13}C NMR spectrum, the two benzene rings must be *p*-disubstituted. The singlet at δ 5.05 may be assigned to a =CH or to a methine (CH) group strongly deshielded by three substituents.

The proton-decoupled ^{13}C NMR spectrum (Fig. 3.4.4, top) consists of six signals corresponding to three CH (δ 131.3, 128.7, and 69.6) and three quaternary C as follows from the DEPT135 measurement (Fig. 3.4.4, bottom). The chemical shift of δ 69.6 indicates that the signal corresponds to an sp^3-hybridized methine group. Two of the quaternary carbon atoms (δ 136.2 and 134.3) are part of an aromatic system, whereas the signal at δ 100.8 may arise from an sp^2 or a deshielded sp^3 carbon atom.

The two-dimensional HSQC spectrum (Fig. 3.4.5) provides the ^1H and ^{13}C chemical shifts of the CH moieties. The 5.05/69.6 cross peak proves the sp^3 character of this methine group.

As to the two- and three-bond ^1H–^{13}C connectivities, they can be obtained from the two-dimensional HMBC spectrum (Fig. 3.4.6). The methine proton gives cross peaks with two quaternary carbons (5.05/100.8 and 136.2) and with one set of the aromatic =CH carbon atoms (5.05/131.3) proving their assignment to C-2, C-1' and C-2',6', respectively. Differentiation of the aromatic =CH moieties is also clear from the δ 7.53/69.6 cross peak.

The mass spectrum (Fig. 3.4.1) terminates with a peak cluster around m/z 354, with differences of 2 u between individual maxima within the group. Such a situation is always indicative of the isotope pattern caused by the simultaneous presence of several chlorine and/or bromine atoms. The intensity distribution within the group together with a sequence of mass differences of 35, 36 and 70 u with the appropriate changes in the intensity pattern prove that five chlorine atoms are present.

The loss of three of these chlorine atoms must be involved in forming the dominant base peak at m/z 235. Since the mass difference with respect to the molecular ion (by definition, m/z 352 if the last group of peaks represents the molecular mass) is 117, the neutral moiety lost must be a trichloromethyl group with an easily broken bond. The ^{13}C-isotope peak at m/z 236 has a relative intensity of 15%, which means that up to 13 carbon atoms can be present in the fragment of m/z 235. This nicely fits the presence of two aromatic rings, which are necessary anyway to account for eight protons in the ^1H NMR spectrum and which must be attached to the methine group. The loss of two chlorine atoms from the base-peak fragment produces the next most prominent fragment at m/z 165, which agrees with the expected elemental composition of $C_{13}H_9$. The molecular formula of the compound, thus, corresponds to $C_{14}H_9Cl_5$ with eight double bond equivalents.

3.4.2   Structure Assembly

Two *p*-chlorophenyl rings, one trichloromethyl group, and a methine carbon atom can only be arranged in one way to give 1,1-bis(4-chlorophenyl)-2,2,2-trichloroethane (DDT):

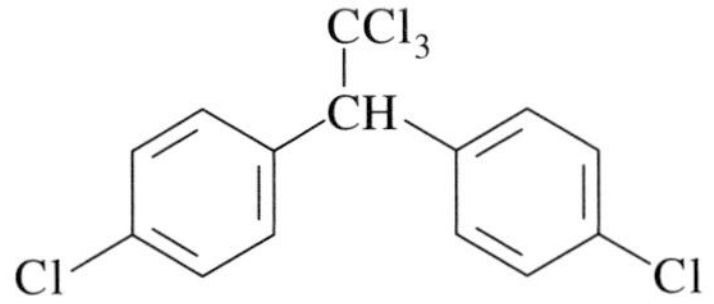

Check with Assemble 2.1

For molecules with conformational flexibility, one single conformation of high symmetry can be used to judge, which nuclei and which coupling paths are equivalent. Moreover, if there exist groups with fast rotation around a bond that coincides with a symmetry axis of this group, atoms and coupling paths permutated by the rotation will be equivalent (cf. Chapter 4.3). In the present case, a symmetry plane defined by the central H–C–C moiety makes the two benzene rings equivalent. Due to their fast rotation around the arC–CH bond, the protons and carbon atoms in *ortho* and in *meta* position become equivalent. Therefore, only 6 signals are expected in the ^{13}C NMR spectrum.

Assemble 2.1 finds eight structures on the basis of the molecular formula with two *p*-substituted benzene rings. Only two of them are generated if, additionally, the number of signals in the ^{13}C NMR spectrum is set to 6.

3.4.3   Comments

3.4.3.1   Mass Spectrum

Except for the base peak, most fragments arise by consecutive losses of Cl, HCl, and Cl_2.

3.4.3.2   Infrared Spectrum

The type of aromatic substitution can be deduced from the presence of only two combination bands at 1900 and 1780 cm^{-1} and their intensity ratio. The prominent bands at 1095 and 1015 cm^{-1} result from combined C–Cl and C–C stretching vibrations, whereby the one at 1095 cm^{-1} is characteristic of a *p*-substituted chlorophenyl group.

3.4.3.3 ^1H and ^{13}C NMR Spectra

The estimation of chemical shifts is often helpful for assigning them or for excluding alternative structures. The mean deviations between estimated and measured shifts is of the order of 2–3 and 0.2–0.3 ppm for ^{13}C and ^1H NMR, respectively.

Check with NMRPrediction

The ^{1}H and ^{13}C chemical shifts can be estimated with the program NMRPrediction. Open the program and draw the structure. Select "Estimate ^{1}H NMR" or "Estimate ^{13}C NMR" from the "Estimate" menu. Detailed information on the parameters used is obtained from the protocol (see "Show Protocol" of the "Estimate" menu).

It is possible to record HSQC (Fig. 3.4.5) and HMBC spectra (Fig. 3.4.6) in reasonable time. Their combined application provides unambiguous assignments of all the ^{1}H and ^{13}C NMR signals.

3.4.3.4   Presentation of NMR Data

^{1}H NMR (500 MHz, CDCl$_3$): δ = 7.53 (dm, J = 8.6 Hz, 4H, H-2',6'), 7.34 (dm, J = 8.6 Hz, 4H, H-3',5'), 5.05 (s, 1H, H-1). ^{13}C NMR (125 MHz, CDCl$_3$): δ = 136.2 (C-1'), 134.3 (C-4'), 131.3 (C-2',6'), 128.7 (C-3',5'), 100.8 (C-2), 69.6 (C-1).

3.5 Problem 5

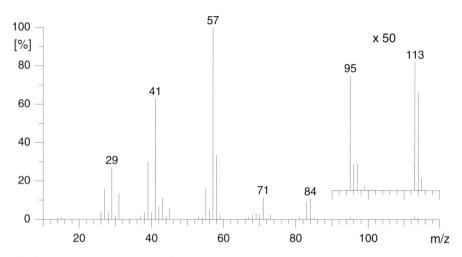

Fig. 3.5.1: Mass spectrum, EI, 70 eV

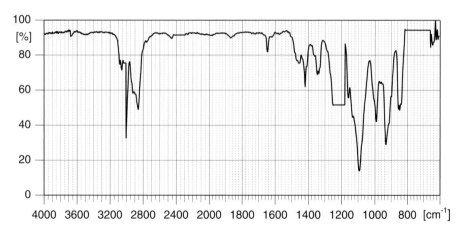

Fig. 3.5.2: IR spectrum, solvent CHCl₃, cell thickness 0.2 mm

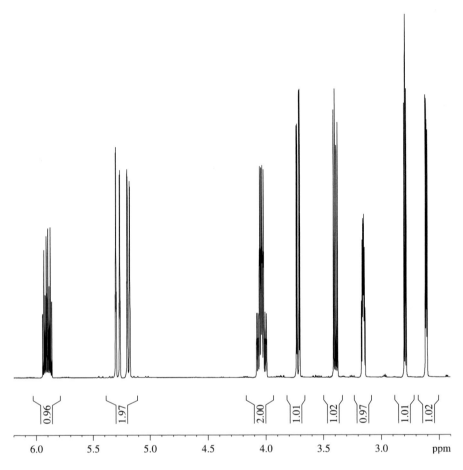

Fig. 3.5.3: ^{1}H NMR spectrum, 500 MHz, solvent CDCl$_3$

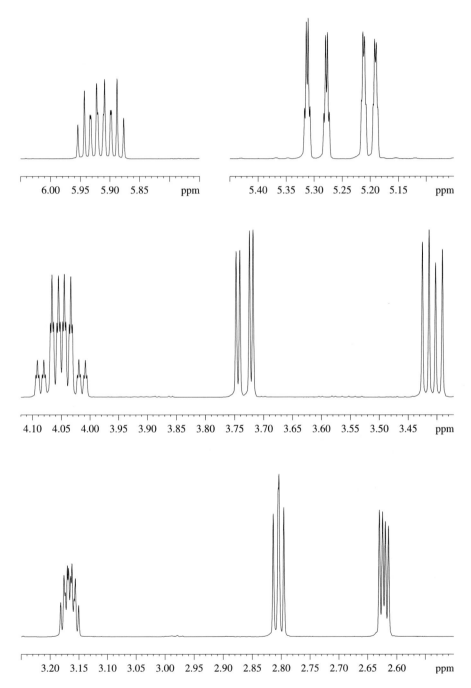

Fig. 3.5.4: Expanded ¹H NMR spectrum, 500 MHz, solvent CDCl₃

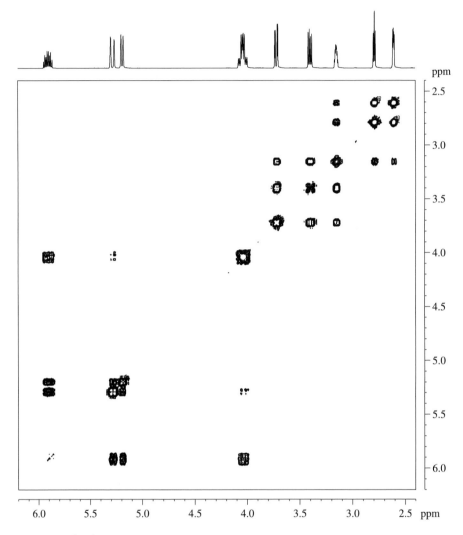

Fig. 3.5.5: ^{1}H,^{1}H COSY spectrum, 500 MHz, solvent CDCl$_3$

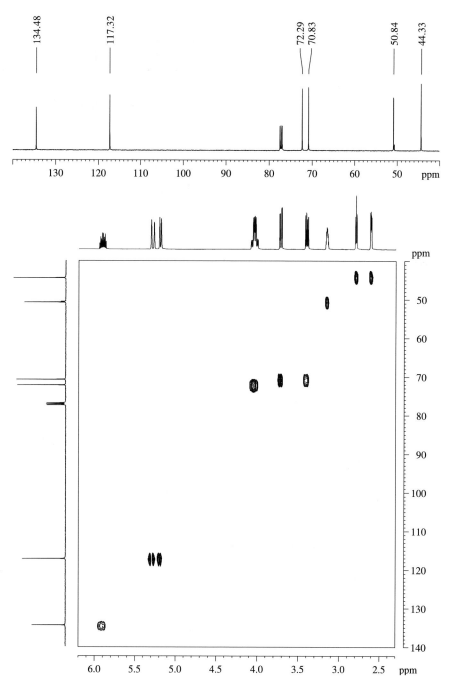

Fig. 3.5.6: Top: ^{13}C NMR spectrum, 125 MHz, solvent CDCl$_3$. Bottom: ^{13}C,^{1}H HSQC spectrum

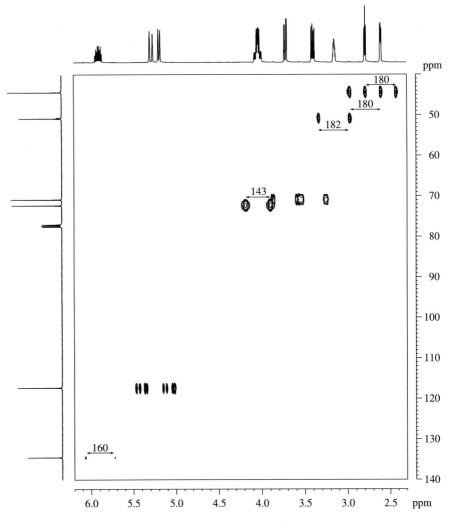

Fig. 3.5.7: HSQC spectrum with $^1J_{C,H}$, solvent CDCl$_3$

3.5.1 Elemental Composition and Structural Features

The ^{1}H NMR spectrum (Fig. 3.5.3) shows eight signals in the range of δ 6.0–2.0 with relative intensities of 1:2:2:1:1:1:1:1 (from left to right), which gives a total of 10 protons.

According to the ^{13}C NMR spectrum (Fig. 3.5.6, top), six non-isochronous carbon atoms are present in the molecule. The signals at δ 134.5 and 117.3 indicate a C=C double bond. Considering the HSQC map (Fig. 3.5.6, bottom), it belongs to a vinyl group with the =CH at δ 134.5 and the =CH$_2$ at δ 117.3. Three signals in the sp^3-carbon region (δ 72.3, 70.8, and 44.3) are each correlated with two protons (CH$_2$), and the signal at δ 50.8 with one proton (CH). This HSQC spectrum further indicates that all protons are bonded to carbon atoms. The chemical shifts around δ 70 for the sp^3 methylene groups demand the presence of a neighboring oxygen substituent.

In the IR spectrum (Fig. 3.5.2), we find the vinyl group confirmed: stretching vibrations at 3060 and 1640 cm^{-1} for C–H and C=C, respectively, and C–H out-of-plane deformation vibrations below 1000 cm^{-1}. Furthermore, there is a strong band at 1090 cm^{-1}, which might be a C–O stretching vibration of an ether.

According to the mass spectrum (Fig. 3.5.1), the molecular mass could be 113 or 114. In the former case, m/z 114 is too strong for a ^{13}C-isotope peak, so it must predominantly originate from the protonated molecular ion. If, on the other hand, m/z 114 corresponds to the molecular ion, m/z 113 is formed by the loss of a hydrogen radical. The first alternative can be excluded because we have an even number of hydrogen atoms and there are no indications for halogens. An odd nominal molecular mass implies an odd number of nitrogen atoms in the molecule which, in turn, requires an odd number of odd-valence elements.

The molecule easily fragments in two halves of equal mass (m/z 57). The elements found so far (C_6H_{10}) correspond to a mass of 82 u, leaving 32 u to complete the molecule. This is most readily accomplished by adding two oxygen atoms or one sulfur atom. However, the mass spectrum shows no indication for sulfur. Since the ^{13}C chemical shifts around δ 70 for the aliphatic methylene groups demand the presence of a neighboring oxygen, the elemental composition becomes $C_6H_{10}O_2$, corresponding to two double bond equivalents. As there is only one C=C and no C=O double bond (see the information from the IR and ^{13}C NMR spectra), the molecule must contain one ring.

3.5.2 Structure Assembly

As indicated by the chemical shifts in both NMR spectra, the vinyl group is not bonded to an oxygen atom since for a CH$_2$=CH–O– group, ^{1}H shift values of ca. δ 6.4, 4.0, and 3.9 and ^{13}C chemical shift values of δ 150 and 85 would be expected.

Based on the molecular formula alone, Assemble 2.1 generates 4 587 isomers if improbable fragments are excluded. This number is further reduced to 36 if one ring is required as well as the substructures found above, i.e., a vinyl group not bonded to oxygen and two CH$_2$ groups with a neighboring O.

Check with Assemble 2.1

The use of atom tags is recommended for the input of the above-specified substructures. Open Edit Forbidden Fragments and check Forbid All and then Apply. Draw the substructure CH$_2$=CH–R, select the CH group, and open (with the right mouse button) the Atom Tag Dialog Box. Select Neighboring Atom, enter C, and be sure to check "any" for Hybridization and for Bond. Since the CH has already one C as a neighbor, its total number of neighboring C atoms, Min and Max, must each be set to 2 (see Fig. 3.5.8).

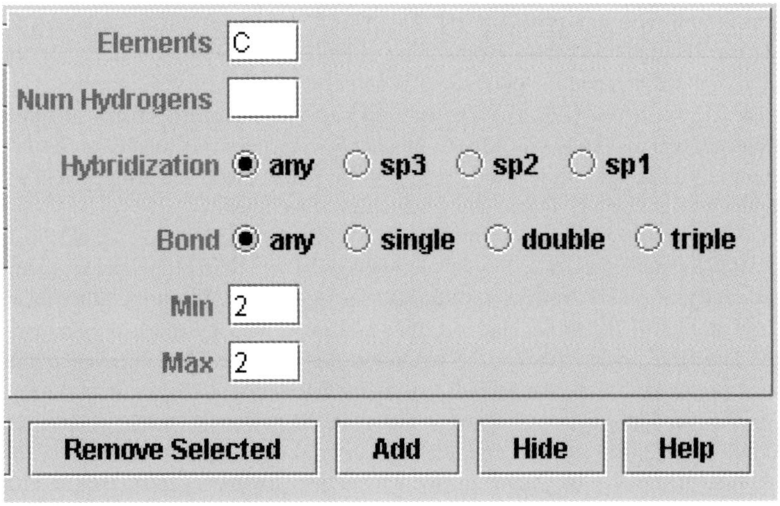

Fig. 3.5.8: Input of atom tags

In the same way, enter the R–CH_2–R group with an atom tag on CH_2 demanding one single-bonded O. In the Assemble Input Window, enter that this substructure must occur twice (Fig. 3.5.9). With this input, Assemble 2.1 produces 33 isomers.

The correct use of atom tags requires sharp reasoning and some practice. If, for example, the input is modified so that only one of the two R–CH_2–R fragments has O as atom tag (i.e., it corresponds to R–CH_2–O–R), 36 isomers are generated instead of 33. The reason is that the atom tag allowing only one O neighbor excludes structures with O–CH_2–O units. Entering twice the substructure CH_2O (without atom tags) would have a different meaning, requiring that each of the two CH_2 groups be bonded to different oxygen atoms.

The proton-proton connectivities across two or three bonds can be obtained from the COSY spectrum (Fig. 3.5.5). The =CH_2 methylene protons at δ 5.30 and 5.20 correlate with each other and with the =CH proton at δ 5.92. The cross peak at δ 5.92/4.05 leads to the neighboring CH_2O group. On the other hand, one of the weak allylic couplings of this CH_2O with the =CH_2 methylene protons leads to a weak cross peak between them. Since no cross peaks connect these CH_2O protons with the remaining ones, this spin coupling network ends here. The group CH_2=CH–CH_2–O- (C_3H_5O, 57 u) constitutes one half of the molecule.

```
┌─────────────────────────────────────────────────────────────────────┐
│ ▓ Assemble 2.1.1  Input of Pb5                            _ □ ☒      │
│ File  Edit  Project  Help                                            │
│ ──────────────────────────────────────────────────────────────────  │
│  Molecular Formula │C6H10O2                                    │     │
│                                                                      │
│  Number of:        min   max    Atoms:              min max          │
│  Rings            │ 1 │ │ 1 │   ┌──────────────────────────┐        │
│  Cycles           │   │ │   │   │                          │        │
│  Double Bonds     │   │ │   │   │                          │        │
│  Triple Bonds     │   │ │   │   │                          │        │
│  H O-CHn, N-CHn   │   │ │   │   Cycle Sizes:      min max          │
│  H cyclopropylic  │   │ │   │   ┌──────────────────────────┐        │
│  H vinyl-aromatic │   │ │   │   │                          │        │
│  C13-NMR Signals  │   │ │   │   │                          │        │
│                                                                      │
│  Forbidden Fragments: 7                                              │
│                                                                      │
│  Fragments: 2                       min    max   overlapping         │
│  ┌──────────────────────────────┐  ┌─────┐                          │
│  │         <C 22>               │  │  1  │        ☐                  │
│  │   C════════R                 │  └─────┘                          │
│  ├──────────────────────────────┤                                   │
│  │         <O sp311>            │  ┌─────┐                          │
│  │        /\                    │  │  2  │        ☐                  │
│  │     R     R                  │  └─────┘                          │
│  └──────────────────────────────┘                                   │
└─────────────────────────────────────────────────────────────────────┘
```

Fig. 3.5.9: The Asssemble 2.1 input window with the fragments

For the other half, having the same elemental composition and one ring, the following possibilities can be considered:

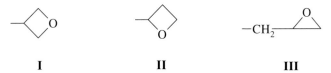

        I   **II**   **III**

Substructure **I** can be excluded because its two methylene carbon atoms would be equivalent by symmetry. Substructure **II** would have to contain an O–CH–O group. Such groups have ^{1}H-chemical shifts around δ 6 and ^{13}C-chemical shifts of δ 90–100.

Since there are no resonances in these NMR ranges, **II** can be excluded with certainty. Moreover, the interpretation of the five signals of this moiety in the COSY spectrum leads to an unambiguous result. The methine proton at δ 3.17 is coupled with both CH$_2$O groups, whereas no coupling is seen between the protons of these two groups. This situation is fulfilled only in substructure **III**. Hence, the structure of the compound in question is:

3.5.3 Comments

3.5.3.1 Mass Spectrum

The mass spectrum nicely reflects the general rule that upon cleavage of bonds to hetero atoms, the positive charge preferentially stays with the carbon atom due to effects of electronegativity. The two most intense peaks at m/z 57 and 41 are the result of fragmentation at the ether oxygen atom, m/z 58 is due to the allyl alcohol cation obtained by hydrogen rearrangement and, presumably, is the precursor of m/z 31. The signals at m/z 31, 45, 58, and 73 could be used as direct evidence for fragments containing oxygen.

3.5.3.2 Infrared Spectrum

On the whole, C–H stretching frequencies between 3100 and 3000 cm^{-1} are observed only for hydrogen atoms bonded either to an sp^2-hybridized carbon atom or to carbon-bearing halogen atom(s). In addition, three-membered rings also exhibit C–H stretching frequencies in this range. For aromatic moieties, the respective absorption bands are generally quite weak and the frequencies are near the lower limit. Medium intensity bands at the upper frequency limit are observed for methylene groups that are terminal or part of a three-membered ring, and sometimes for hydrogen atoms in five-membered heteroaromatic systems. In the present case, the absorption at 3080 cm^{-1} is assigned to the methylene group in the oxirane ring and to the methylene of the vinyl group.

3.5.3.3 ^{1}H NMR Spectrum

On the basis of the difference between *cis* and a *trans* coupling constants of ca. 10 and 17 Hz, respectively (vicinal couplings with the proton at δ 5.92, cf. Fig. 3.5.4), it is possible to distinguish between the vinyl methylene protons. One would expect two doublets if no other couplings were present. However, the signals are further split by a small geminal coupling (0–4 Hz) and by *cisoid* and *transoid* allylic coupling. In the present case, all three coupling constants are similar so that the lines are split into quasi-quartets. The ddt (doublet of doublet of triplet) multiplicity of the =CH signal at δ 5.92 is in accord with the number of vicinal coupling partners. At first glance, the pattern of the =CH$_2$ signal at δ 5.30/5.20 is rather complicated, indicating that the methylene protons are anisochronous. Since the molecule is chiral, the methylene protons are diastereotopic and, thus, in principle anisochronous. As a consequence, the geminal coupling shows up and leads to a higher-order spectrum which, however, can often be understood by first-order rules. Notwithstanding, borderline cases between first- and higher-order spectra frequently contain pitfalls so that, as a check, it is recommended to calculate the spin system. The same holds true for the following remarks: Vinyl groups often lead to *ABC*-type spectra which are close to *ABX* cases. The *X* part of an *ABX* system generally consists of four lines (in some cases, the theoretically possible six lines are observed) positioned symmetrically around the center (chemical shift). The symmetry-equivalent lines have the same intensities. In borderline cases between *ABX* and *ABC* systems, there are still four lines that are symmetrical with respect to their position but no longer symmetrical regarding their intensities. It is important to keep in mind that *ABX* systems lead to higher-order spectra and the line spacings in the *X* part need not be equal to the coupling constants. The basic pattern of the signal at δ 4.05, but not necessarily the coupling constants, can be understood as the overlap of two signals, each split into the doublet (geminal coupling) of a doublet (vicinal coupling) of a triplet (two allylic couplings with the same coupling constant). The *cisoid* and *transoid* allylic coupling constants do not necessarily have the same value. In the present example, they are almost equal.

The two protons of the other OCH_2 group (δ 3.73 and 3.41) show characteristically different chemical shifts and coupling constants. The protons and carbon atoms in three-membered rings are more strongly shielded than the corresponding ones in other cyclic or acyclic structures. In three-membered rings, the absolute value of the geminal coupling constants is small (^2J ca. 5 Hz). Considering the general rule that in three-membered rings, in accord with the Karplus equation, $^3J_{cis} > {}^3J_{trans}$, differentiation between the two methylene protons of the epoxide ring is straightforward.

3.5.3.4 ^{13}C NMR Spectrum

The signals of sp^2- and sp^3-hybridized carbon atoms are easily differentiated on the basis of their chemical shift values. The signal assignment of the attached CH pairs follows from the HSQC spectrum (Fig. 3.5.6, bottom). Performing the same experiment without broadband ^{13}C decoupling yields the ^{13}C-coupled HSQC spectrum (Fig. 3.5.7), in which the cross peaks are split due to the $^1J_{C,H}$ coupling. These coupling constants can be directly read from the spectrum. Owing to the effect of strong ring strain in epoxides, the value of $^1J_{C,H}$ is ca. 80 Hz, which is of diagnostic value for this group. For the OCH_2 groups in chains or in rings without strain, a value of ca. 145 Hz is expected and for =CH or aromatic CH bonds ca. 160 Hz. As in the ^1H NMR spectrum, the signals of epoxide carbon atoms are shifted upfield relative to the corresponding carbon atoms in other cyclic or acyclic ethers.

3.5.3.5 Presentation of NMR Data (500 resp. 125 MHz, CDCl$_3$, δ)

Assignment	^{1}H (J)	^{13}C ($^1J_{CH}$)
=CH$_2$	5.30, dq (17.2, 1.6 Hz),	117.3
	5.20, dd (10.3, 1.6 Hz)	
=CH	5.92, ddt (17.2, 10.3, 5.5 Hz)	134.5 (160 Hz)
OCH$_2$	4.05, m	72.3 (143 Hz)
OCH$_2$	3.73, dd (11.5, 3.1 Hz)	70.8
	3.41, dd (11.5, 5.9 Hz)	
CH	3.17, m	50.8 (182 Hz)
CH$_{cis}$	2.80, dd (5.0, 4.7 Hz)	
CH$_{trans}$	2.62, dd (5.0, 2.7 Hz)	44.3 (180 Hz)

3.6 Problem 6

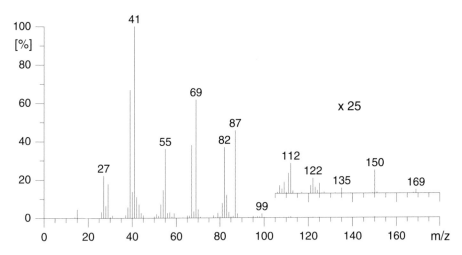

Fig. 3.6.1: Mass spectrum, GC-MS, EI, 70 eV

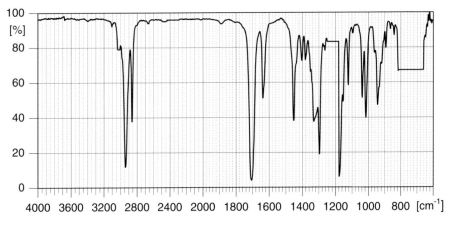

Fig. 3.6.2: IR spectrum, solvent CHCl₃, cell thickness 0.2 mm

UV spectrum (solvent ethanol): λ_{max} = 208 nm, log ε = 3.9

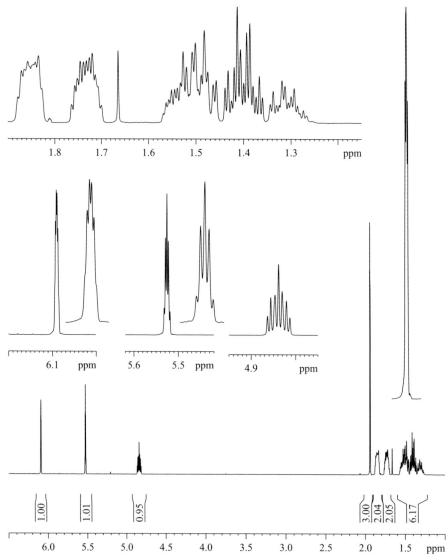

Fig. 3.6.3: ^{1}H NMR spectrum, 500 MHz, solvent CDCl$_3$

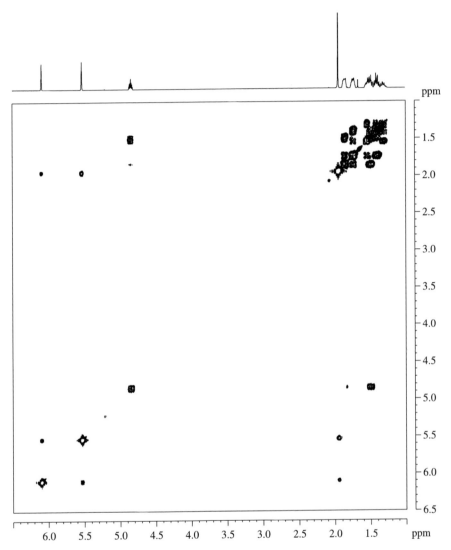

Fig. 3.6.4: ^{1}H,^{1}H COSY spectrum, 500 MHz, solvent CDCl$_3$

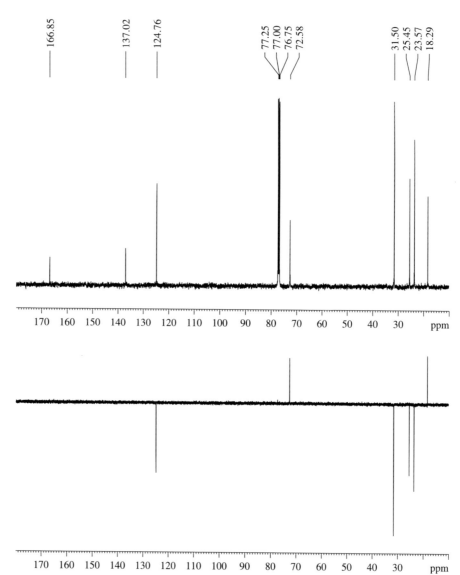

Fig. 3.6.5: ^{13}C NMR spectra, 125 MHz, solvent CDCl$_3$. Top: proton-decoupled; bottom: DEPT135

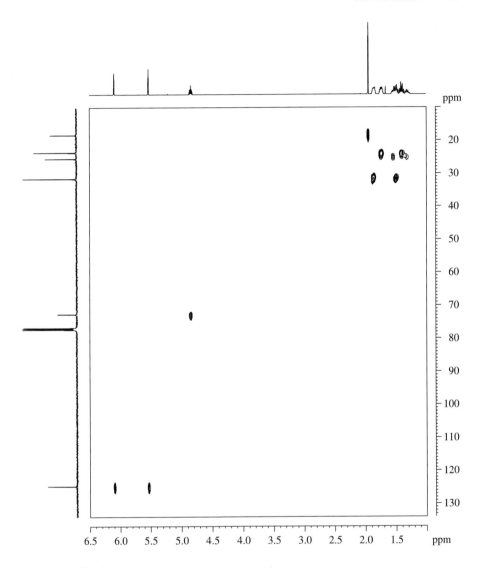

Fig. 3.6.6: ^{13}C,^{1}H HSQC spectrum, solvent CDCl$_3$

3.6.1   Elemental Composition and Structural Features

The IR spectrum (Fig. 3.6.2) exhibits a strong absorption at 1710 cm^{-1} in the carbonyl region and a band at 1640 cm^{-1}, which can be assigned to a carbon-carbon double bond. Because of the relatively high intensity of this band, we expect the double bond either to be linked to a hetero atom, or be *cis*-disubstituted, or have a terminal methylene group.

The ^{1}H NMR spectrum (Fig. 3.6.3) shows seven signals at δ 6.09, 5.53, 4.84, 1.94, 1.85, 1.73, and 1.60–1.25. The integration gives a proton ratio of 1:1:1:3:2:2:6 (from left to right), giving a total of 16 H atoms in the molecule. Judging from their chemical shift values, the first two signals are due to olefinic protons. As they only show very small splittings, they are most likely bonded to the same carbon atom in a terminal vinylidene group, in line with IR spectral evidence.

The ^{13}C NMR spectrum (Fig. 3.6.5, top) presents eight signals; three of them, corresponding to two quaternary C atoms and one methylene, appear in the area of the sp^2 carbons, the other carbon atoms, i.e., one CH, three CH$_2$, and one CH$_3$, are sp^3-hybridized. The signal at δ 166.9 confirms the presence of a carbonyl group. From IR spectral evidence, a ketone would very well fit the observed frequency. However, the chemical shift definitely excludes a ketone because we then would find the carbonyl signal near or above δ 200. This conflict is resolved by assuming an α,β-unsaturated ester.

The two signals at δ 137.0 and 124.8 are assigned to the C atoms forming the carbon-carbon double bond. The line at δ 72.6 corresponds to the postulated methine group, its chemical shift value corroborating the assumption of a neighboring oxygen atom. The signal at the high-field end of the spectrum (δ 18.3) arises from a methyl group. The remaining three methylene signals have rather different intensities and must correspond to 10H; therefore, we have to assume that some lines in the ^{13}C NMR spectrum coincide.

If we now add up all the structural fragments found so far, we get C$_{10}$H$_{16}$O$_2$ with a molecular mass of 168. This is at variance with the mass spectrum (Fig. 3.6.1), which ends with a low intensity peak at m/z 169. The difference is most easily rationalized by assuming protonation of the molecular ion, which is not unusual with esters. The spectrum is of the aliphatic unsaturated type (m/z 27, 41, 55, 69), with one prominent even-mass fragment ion at m/z 82. The fragments at m/z 150 and 135 arise from the consecutive loss of 18 (water) and 15 u (methyl radical) from the (unprotonated) molecular ion.

3.6.2   Structure Assembly

We have, so far, identified the following fragments:

1	$-CH_3$
2	$-CH_2-$
2	$-CH_2-$
1	$-CH_2-$

Check with Assemble 2.1

With the molecular formula, C$_{10}$H$_{16}$O$_2$, Assemble 2.1 generates 4 673 363 isomers. Only 7 of them are possible with the substructures shown above and the required molecular symmetry. Check it with Assemble 2.1!

When using Assemble 2.1, one should keep in mind that global information, such as the number of CH_2 groups or the number of specified neighbors (atom tags), always means the *total* number, i.e., including those explicitly shown in a substructure. Thus, a total of 6 CH_2 groups must be specified in the Atoms Window (if no hybridization is considered) and not 5 as shown above to the right.

In addition to the above left fragment with three free valences, we obviously have only one terminal function, namely the methyl group. This indicates that a ring must be present. When calculating the double bond equivalents for the elemental composition found, we indeed obtain three. The methyl group cannot be placed at the methine group since it would give rise to a doublet in the ^1H NMR spectrum. We, therefore, locate it at the double bond and then expect a singlet around δ 2 broadened or split by long-range coupling with the alkene protons. The signal of these protons, in turn, is split up with the same small coupling constant. The actual spectrum shows a doublet of doublets (dd) for the methyl group, which is in line with our hypothesis.

We now have to place the five methylene groups on the CH so as to (a) form a ring and (b) create two pairs of isochronous carbon atoms. There is only one possible solution, which, therefore, represents the constitution of the unknown compound in question:

3.6.3 Comments

3.6.3.1 Mass Spectrum

The repeatability (same instrument, operator, and day) of mass spectra is quite high. However, the reproducibility (different instruments and operators) can be surprisingly low. Intensities may change considerably in particular if the peak is produced by a bimolecular reaction, e.g., by protonation of the molecular ion. This is illustrated by the mass spectrum shown in Fig. 3.6.7 of the sample recorded on a magnetic sector instrument rather than on a GC-MS combination apparatus with a quadrupole mass filter as for Fig. 3.6.1.

Both mass spectra show two important fragmentation paths common to esters, namely a double hydrogen rearrangement to give the protonated acid of m/z 87, which then successively loses water and carbonyl (m/z 87 → 69 → 41), and an acid elimination from the molecular ion to yield the cyclohexene radical cation of m/z 82, which subsequently loses (typically) methyl to give the fragment C_5H_7 (m/z 67). The remaining major features of the mass spectra are accounted for by the formation of cyclohexyl ion (m/z 83) and by the loss of ethylene from both m/z 83 and 82 to yield m/z 55 and 54, respectively.

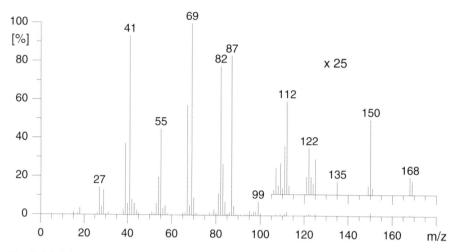

Fig. 3.6.7: Mass spectrum, EI, 70 eV

3.6.3.2 Infrared Spectrum

The intensity of C=C stretching absorption bands varies widely. In symmetrically *trans*- or in symmetrically tetrasubstituted double bonds, the C=C stretching frequency becomes inactive in the IR spectrum (it is, however, strong in the Raman spectrum). Even when the two or four substituents are not exactly alike, and in trisubstituted double bonds, the intensity can be quite low. Monosubstituted, *geminally* and *cis*-disubstituted double bonds generally exhibit a band of medium intensity. The strongest absorptions are observed with double bonds directly linked to an oxygen or nitrogen atom or in conjugation with a carbonyl group.

3.6.3.3 ^{1}H NMR Spectrum

The coupling patterns of the =CH$_2$ signals at δ 6.09 (doublet of quartet, dq) and 5.53 (quintet) indicate that in addition to small geminal coupling ($^2J \approx 1.7$ Hz) long-range couplings are present. From the two-dimensional COSY spectrum (Fig. 3.6.4), the coupling partners are easily identified as the =CH$_2$ and CH$_3$ protons. The absolute magnitudes of allylic coupling constants, which vary between 0 and 3.5 Hz, are mainly influenced by stereochemistry. *Cisoid* and *transoid* allylic coupling constants, on the whole, have different values. If the double bond is not exocyclic, *cisoid* coupling constants are generally somewhat larger than *transoid* ones. In the present spectrum, the quintet multiplicity of the signal at δ 5.53 is explained by the fact that the geminal and allylic coupling constants are approximately equal. The barely resolved signal at δ 6.09 exhibits smaller allylic couplings (≈ 1 Hz); therefore, it can be assigned to the proton in position *E* (*trans*) relative to the methyl group and the signal at δ 5.53 to that in position *Z* (*cis*).

In monosubstituted cyclohexanes, the substituent prefers the equatorial position and the methine hydrogen is axial in the predominantly occurring conformation. Thus, the methine signal at δ 4.84 exhibits two diaxial (large, ≈ 10 Hz) and two diequatorial (small, ≈ 4 Hz) couplings and has a triplet of triplet (tt) structure in the first-order spectrum. Since the coupling partners are part of a higher-order spectrum, splitting patterns and line widths are influenced by higher-order effects. In the present example, however, we find the expected (partly overlapping) triplet of triplets.

In the COSY spectrum, the methine proton shows two cross peaks at δ 1.50 (2H) and 1.85 (2H), which correspond to the neighboring axial and equatorial methylene protons, respectively. The higher intensity of the former and the smaller intensity of the latter cross peak reflect the different magnitudes of the vicinal couplings ($J_{ax,ax} > J_{ax,eq}$), whereas the cross peaks arising from geminal couplings have similar intensities as those of the diaxially coupled protons. The signals of the neighboring methylene groups can be assigned at δ 1.73 (2H) and 1.40 (2H), respectively. Despite of strong overlapping, the chemical shifts of the remaining methylene protons of the cyclohexyl ring (δ 1.55, 1H and 1.32, 1H) can also be identified. It is worth mentioning that for protons of a cyclohexyl moiety, $δ_{ax} < δ_{eq}$ holds as a general rule.

The assignment of the overlapping proton signals of the cyclohexyl moiety is also corroborated by the HSQC spectrum (Fig. 3.6.6). Often, the cross peaks of such a spectrum allow the ^{1}H chemical shifts to be read with acceptable accuracy even in cases of strongly overlapping ^{1}H NMR signals.

3.6.3.4   Presentation of NMR Data (500 resp. 125 MHz, CDCl₃, δ)

Assignment	^{1}H (J)	^{13}C
=CH₂ (E)	6.09, dq (1.7, ≈ 1 Hz), 1H,	124.8
(Z)	5.53, quintet (1.7 Hz), 1H	
=C	–	137.0
CH₃	1.94, dd (1.7, ≈ 1 Hz), 3H	18.3
OCH	4.84, tt (9, 4 Hz), 1H	72.6
CH₂	1.85, m, 2H	31.5
	1.50, m, 2H	
CH₂	1.73, m, 2H	23.6
	1.40, m, 2H	
CH₂	1.55, m, 1H	25.5
	1.32, m, 1H	
C=O	–	166.9

3.7 Problem 7

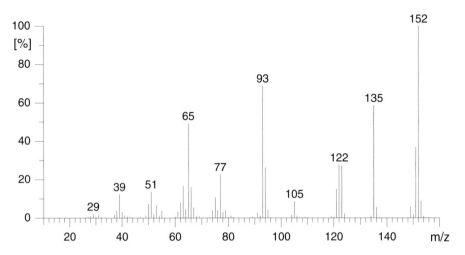

Fig. 3.7.1: Mass spectrum, EI, 70 eV

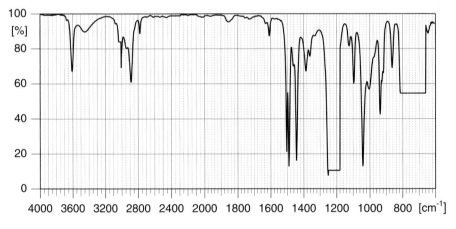

Fig. 3.7.2: IR spectrum, solvent CHCl₃, cell thickness 0.2 mm

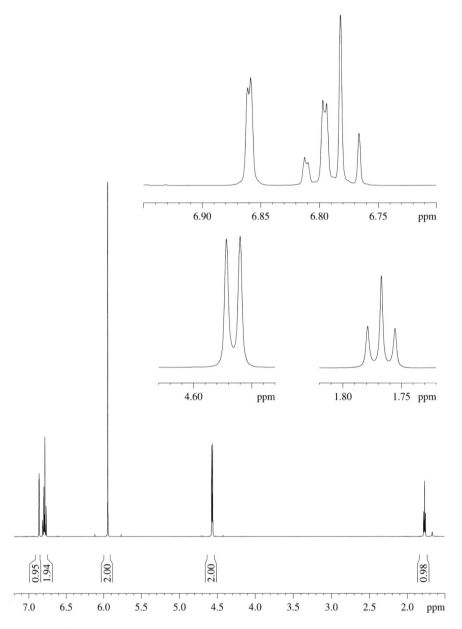

Fig. 3.7.3: ^{1}H NMR spectrum, 500 MHz, solvent CDCl$_3$

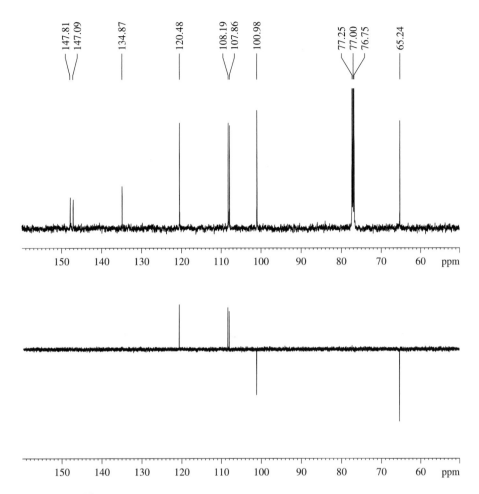

Fig. 3.7.4: ^{13}C NMR spectra, 125 MHz, solvent CDCl$_3$. Top: proton-decoupled; bottom: DEPT135

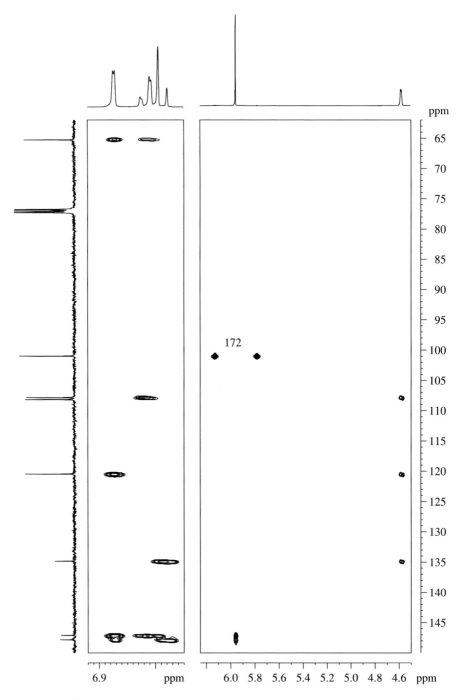

Fig. 3.7.5: $^{13}C,^{1}H$ HMBC spectrum, solvent CDCl$_3$

3.7.1 Elemental Composition and Structural Features

The IR spectrum (Fig. 3.7.2) clearly shows the characteristic bands of a hydroxyl group at 3600 (OH stretching vibration free) and 3450 cm^{-1} (OH stretching vibration associated). These absorptions cannot be assigned to NH$_{(2)}$ because NH$_2$ groups give rise to two bands of equal line width and NH groups never absorb at frequencies above 3500 cm^{-1}.

We further note a band at 2780 cm^{-1} in the CH stretching region. Such CH absorption bands at the lower end of the standard range are expected for methyl groups and, sometimes, methylene groups bonded to nitrogen in amines or to oxygen in ethers as well as for aldehydes. Furthermore, in alicyclic systems, we often find weak combination bands in this region. The carbonyl region is empty. An aromatic moiety is indicated by the bands at 1600 and 1500 cm^{-1}.

The mass spectrum (Fig. 3.7.1) is dominated by the peak at m/z 152, most probably corresponding to the molecular ion. The aromatic ion series (m/z 39, 51, 65, 77) accounts for all main peaks in the lower mass region. However, instead of the expected signal at m/z 91 or 92, we find an intense peak at m/z 93. This could be indicative of a hetero atom in the aromatic system or attached to it.

The ^{1}H NMR spectrum (Fig. 3.7.3) contains signals in the aromatic region (δ 6.87 to 6.76), a singlet at δ 5.94, a doublet at δ 4.57, and a triplet at δ 1.77. The integration gives a hydrogen ratio of 3:2:2:1 (from left), summing up to 8 H atoms. The proton-decoupled ^{13}C NMR spectrum (Fig. 3.7.4, top) shows eight signals. On the basis of the chemical shift values and low intensities and from the DEPT135 spectrum (Fig. 3.7.4, bottom), the signals at δ 147.8, 147.1, and 134.9 can be assigned to quaternary sp^2 carbons. The lines at δ 120.5, 108.2, and 107.9 correspond to =CH and those at δ 101.0 and 65.2 to CH$_2$ groups. These two groups have ^{1}H shifts of δ 5.94 and 4.57, respectively. The single proton at δ 1.77, which obviously is not bound to a carbon atom, can be assigned to the OH found in the IR spectrum. On the basis of its vicinal coupling, it must be attached to the CH$_2$ of δ 4.57. On the other hand, the ^{1}H and ^{13}C chemical shifts, respectively, at δ 5.94 (intensity, 2H) and δ 101.0 (CH$_2$) may normally call for a terminal methylidene group. However, there is no evidence at all for a carbon-carbon double bond in the IR spectrum. Furthermore, the appearance of the methylene proton signal as a sharp singlet requires equal chemical shift values for both protons, a rather unlikely coincidence for methylene protons in a molecule without a high degree of symmetry. A rough comparison of the approximate molecular mass due to the atoms already identified (C$_8$H$_8$O, 120 u) with the actual molecular mass of M$_r$ = 152 indicates the presence of another two oxygen atoms and, therefore, excludes a high degree of symmetry. Considering the strong deshielding effect of oxygen substituents, the chemical shift at δ 101.0 is in accordance with a methylenedioxy group. This is further corroborated by the weak, albeit distinct IR band at 2780 cm^{-1}. From the molecular formula, C$_8$H$_8$O$_3$, we calculate five double bond equivalents, hence, the six sp^2 carbon signals (three quaternary C atoms and three =CH) are in line with a benzene ring carrying three substituents. Thus, we have C$_6$H$_3$, –OCH$_2$O–, and –CH$_2$OH as fragments and are now left with the problem of connecting them.

3.7.2   Structure Assembly

There are three possibilities for placing any three substituents on a benzene ring, namely in the 1,2,3-, 1,2,4-, or 1,3,5-positions. The last one may be excluded here since the methylenedioxy group occupies two adjacent positions on the benzene ring. Thus, only two possibilities remain:

I   II

The IR and mass spectra give no reliable information allowing to decide between **I** and **II**. However, the *ABX*-type coupling pattern of the aromatic protons clearly shows that one proton (δ 6.86) has no neighbor in *ortho* position and the other two must be *ortho* to each other (see Comments). Further proof of structure **II** is provided by the two- and three-bond C,H correlations from the HMBC measurement (Fig. 3.7.5).

> **Check with Assemble 2.1**
>
> Based on the molecular formula $C_8H_8O_3$ alone, Assemble 2.1 would generate 6 333 319 structures. However, a manageable number of solutions will be obtained even using only part of the information gained above. By entering the molecular formula as well as the information that a triply substituted benzene ring and an OH group are present and by forbidding improbable structures, 50 isomers are generated. This number is further reduced to 38 if, additionally, the presence of neighboring O is demanded for the two CH_2 groups. Finally, by requiring that the OH group must be attached to an sp^3-hybridized C atom, only 16 structures remain.

3.7.3   Comments

3.7.3.1   Mass Spectrum

Loss of formaldehyde (152 $\rightarrow$ 122, Δm = 30) followed by decarbonylation (122 $\rightarrow$ 94, Δm = 28) is characteristic of methylenedioxy groups on aromatic rings. Degradation of the hydroxymethyl group in benzyl alcohols typically follows the sequence of loss of a hydrogen radical and decarbonylation (m/z 152 $\rightarrow$ 151 $\rightarrow$ 123). These fragmentation paths have been confirmed by independent experiments.

3.7.3.2   Infrared Spectrum

In general, CH stretching vibrations around 3000 cm^{-1} are of low diagnostic value. There are, however, some partial structures that exhibit uncommon bands in this region. On the lower limit of the standard range, i.e., below 2850 cm^{-1}, we find absorption bands for methoxyl groups in ethers (but not in esters). Similar frequencies are found for methyl groups and often also for methylene groups bonded to a nitrogen atom in amines (but not in amides). Furthermore, uncommonly low CH stretching frequencies are exhibited by methylenedioxy groups and by the dioxymethine group in acetals. In addition, absorption bands may be found in the same region for aldehydes and cyclohexane moieties.

To use the absence of a C=C stretching vibration band around 1650 cm^{-1} as an argument against the presence of a double bond is generally unreliable because the intensities can be very low. However, if the double bond is asymmetrically substituted so that the dipole moment changes during vibration, at least a band of medium intensity is expected. Thus, in the present case, it is valid to exclude a terminal methylene group.

3.7.3.3 ^{1}H NMR and ^{13}C,^{1}H HMBC Spectra

The chemical shift and width of signals of hydroxyl protons highly depend on solvent, temperature, and impurities containing exchangeable protons. Because of different possible influences, the shift values in nonpolar solvents are not characteristic of the molecular environment. The signal is often slightly broadened. The chemical shift of hydroxyl protons in aliphatic alcohols without strong hydrogen bonding is generally below δ 5, mostly in the range of δ 1.0–2.5 ppm.

The coupling pattern of the three aromatic proton signals enables the simple decision between the above structures **I** and **II**. Only **I** has a proton with two *ortho* couplings (ca. 8 Hz) as well as a proton showing *meta* coupling (ca. 2 Hz) without *ortho* couplings. *Para* couplings are usually not resolved under standard experimental conditions. The spectrum of the aromatic protons is of higher order but can be interpreted under a first-order approach. Thus, the signal at δ 6.86 exhibits only a *meta* coupling of 1.7 Hz, that at δ 6.80 shows a dd splitting (7.9, 1.7 Hz), whereas the line at δ 6.77 is a doublet with an *ortho* coupling of 7.9 Hz. Due to the strong coupling between the last two protons, one can observe a special intensity distortion of the multiplets, sometimes named roof effect, with the intensities of the inner part of the signals being higher and those of the outer parts, lower. In the case of higher-order spectra, the splittings of the multiplets are usually not equivalent to the exact values of the coupling constants, and also the chemical shifts taken from the spectra are only approximate. To get the exact coupling and δ values, a spectrum simulation is recommended.

The entire proton-carbon coupling network can be obtained from the HMBC spectrum. The methylene protons of the HOCH$_2$– group give correlations *via* two bonds to C-1 (δ 134.9) and *via* three bonds to C-2 and C-6. On the other hand, the –OCH$_2$O– methylene protons specify C-3 and C-4. A differentiation between them can be achieved considering the fact that for aromatic systems, the vicinal ^3J(C,H) coupling constants are ca. 8 Hz so that strong cross peaks appear in the HMBC spectrum. The coupling constants over two bonds, $^2J_{C,H}$, are in the range of 0–2 Hz so that the corresponding HMBC responses are only weak. The H-5 proton causes responses to C-3 and C-1, whereas H-6 shows cross peaks with C-4 and C-2. In the HMBC spectrum, besides the expected H-2/C-4 and H-2/C-6 cross peaks, a correlation of H-2 to the geminal C-3 atom is also observed. The appearance of this cross peak is due to the substitution of C-3 with an electronegative substituent, which may lead to an enhanced value of the geminal $^2J_{C,H}$ coupling constant. Actually, the HMBC measurements do not select between the geminal and vicinal $J_{C,H}$ couplings. The intensity of an HMBC cross peak has a maximum value if the coupling is equal to the set value (in all examples of this book: 7 Hz), but all correlations will be detected that are in the range of 7 ± 3–4 Hz. The intensities of the cross peaks decrease with increasing deviation of the coupling constant from the set value. The one-bond couplings are suppressed in the HMBC spectrum provided that the timing fits the $^{1}J_{C,H}$ coupling constant (i.e., $\tau = 1/(2J)$). The standard setting of $\tau = 3.45$ ms corresponds to $^{1}J_{C,H} = 145$ Hz. Due to the two oxygen substituents, this value is much higher for the –OCH_2O– methylene group and, therefore, the cross peak appears in the spectrum. In such cases, the exact value of this coupling constant can be read from the spectrum (here: 172 Hz). This provides further support of the assignment of the –OCH_2O– signal.

3.7.3.4 Presentation of NMR Data (500 resp. 125 MHz, CDCl₃, δ)

Assignment	^{1}H (J)	^{13}C	HMBC responses (^{13}C partners)
OCH_2	4.57, d (5.8 Hz)	65.2	C-1, C-2, C-6
OH	1.77, t (5.8 Hz)		
OCH_2O	5.94, s	101.0	C-3, C-4 (one-bond: $^{1}J_{C,H} = 172$ Hz)
C-1	–	134.9	
C-2	6.86, d (1.7 Hz)	107.9	OCH_2, C-4, C-6, C-3 weak
C-3	–	147.8	
C-4	–	147.1	
C-5	6.77, d (7.9 Hz)	108.2	C-1, C-3
C-6	6.80, dd (7.9, 1.7 Hz)	120.5	OCH_2, C-2, C-4

3.8 Problem 8

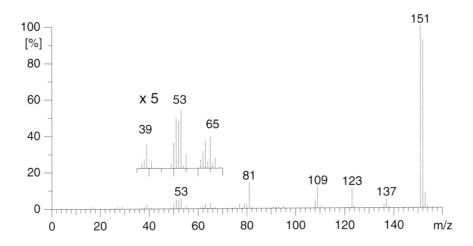

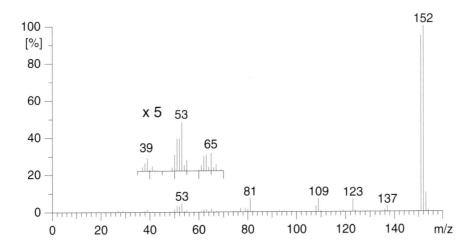

Fig. 3.8.1: Mass spectra, EI, 70 eV. Top: isomer A; bottom, isomer B

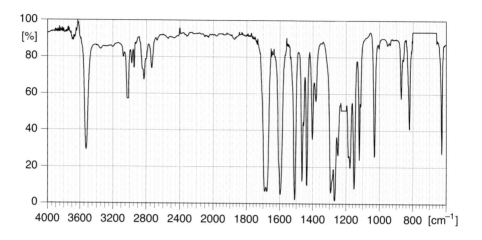

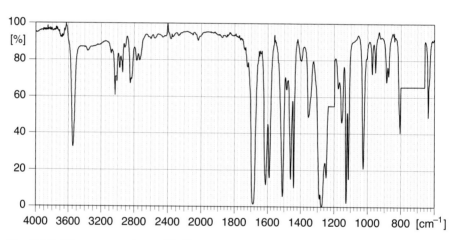

Fig. 3.8.2: IR spectra, solvent CHCl$_3$, cell thickness 0.2 mm. Top: isomer A; bottom, isomer B

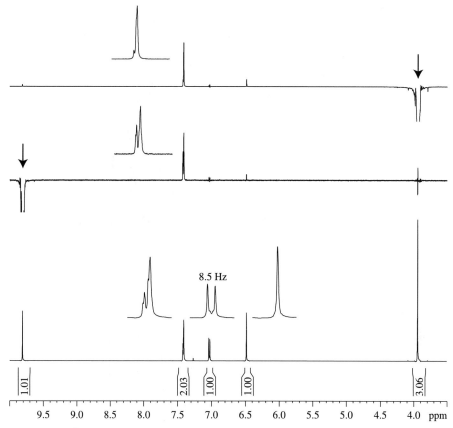

Fig. 3.8.3: ^{1}H NMR spectra of isomer A, 500 MHz, solvent CDCl$_3$. Bottom: conventional spectrum; middle and top: 1D NOESY spectra obtained by irradiating the signals marked with an arrow

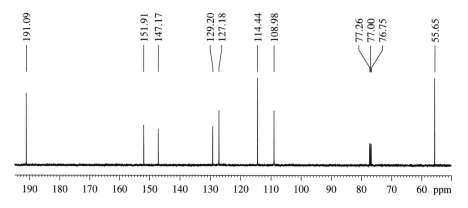

Fig. 3.8.4: ^{13}C NMR spectrum of isomer A, 125 MHz, solvent CDCl$_3$

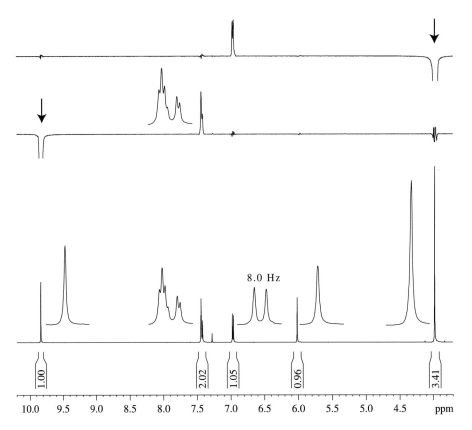

Fig. 3.8.5: ^{1}H NMR spectra of isomer B, 500 MHz, solvent CDCl$_3$. Bottom: conventional spectrum; middle and top: 1D NOESY spectra obtained by irradiating the signals marked with an arrow

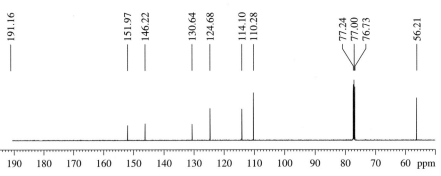

Fig. 3.8.6: ^{13}C NMR spectrum of isomer B, 125 MHz, solvent CDCl$_3$

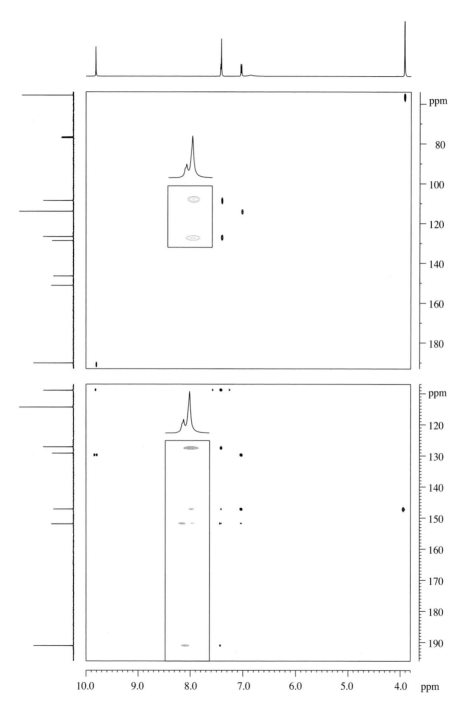

Fig. 3.8.7: ^{13}C,^{1}H HSQC (top) and HMBC (bottom) spectra of isomer A

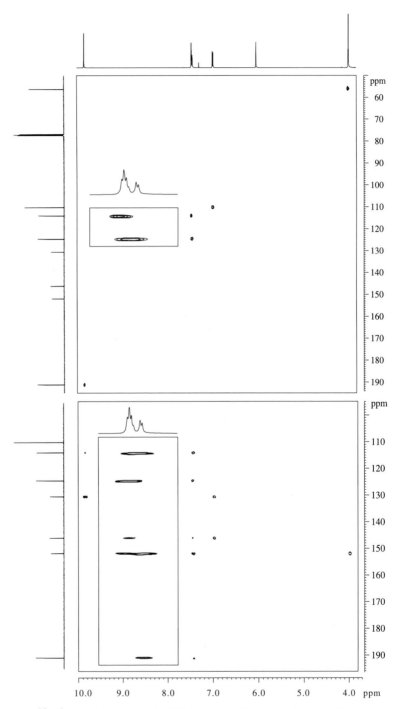

Fig. 3.8.8: $^{13}C,^{1}H$ HSQC (top) and HMBC spectra (bottom) of isomer B

3.8.1 Elemental Composition and Structural Features

The mass spectra of the two isomers A and B (Fig. 3.8.1) do not significantly differ from each other. At first glance, both m/z 151 and 152 may correspond to the molecular ion: In the former case, a significant amount of protonated molecular ion would occur at m/z 152, while in the latter, the molecule would easily loose a H radical. However, the presence of the fragment at m/z 137 allows a clear conclusion owing to its mass difference of 14 u relative to m/z 151. Such a fragmentation is a very unfavorable process and has only been observed in rare special cases, as in the mass spectrum of ketene. The intensity distribution indicates a predominantly aromatic character of the molecule, which is corroborated by the series of highly unsaturated hydrocarbon fragments at m/z 39, 51–53, 63–65, 77–79.

Also the IR spectra (Fig. 3.8.2) of both isomers are very similar. The rather sharp signal at 3530 cm^{-1} indicates an OH group. The C–H stretching vibrations at the lower end of the usual range (2830–2780 cm^{-1}) are characteristic of CH_2 or CH_3 substituted by O or N and those > 3000 cm^{-1} confirm the presence of an aromatic moiety. The latter is further supported by the skeletal vibrations around 1600 and 1500 cm^{-1}. The dominating signal at 1690 cm^{-1} must be due to a C=O group.

The integration of both ^{1}H NMR spectra (Figs. 3.8.3 and 3.8.5) gives a proton ratio of 1:2:1:1:3 (from left) and corresponds to a total of 8 H atoms. The singlet at δ 9.81 (isomer A)/9.83 (isomer B) suggests the existence of an aldehyde group, which is verified by the ^{13}C NMR (Figs. 3.8.4 and 3.8.6) and HSQC spectra (Figs. 3.8.7, top and 3.8.8, top) indicating a CH at δ 191.1/191.2. The signals in the aromatic region near δ 7.4 m (2H) and 7.03/6.97 d (1H, J = 8.5/8.0 Hz) show the presence of a trisubstituted phenyl ring. The singlet at δ 3.94/3.98 of intensity 3H suggests a methoxy group. The remaining signal at δ 6.48/6.01 can be assigned to the OH group found in the IR spectrum since it does not show any correlations to a carbon atom in the HSQC spectra.

The assignment of the eight signals in the ^{13}C NMR spectra is straightforward from the HSQC spectra. The signal at δ 191.1/191.2 is characteristic of HC=O, whereas the peak at δ 55.7/56.2 indicates a CH_3O group. The remaining signals for three =CH and three quaternary C atoms are in accord with a trisubstituted phenyl ring.

3.8.2 Structure Assembly

The elemental composition of the structural fragments found so far, i.e., a trisubstituted benzene ring, –OH, –OCH$_3$, and –CH=O, sums up to $C_8H_8O_3$ corresponding to a mass of 152 u, the molecular mass found above.

Check with Assemble 2.1
With the molecular formula $C_8H_8O_3$ and the fragments identified above (see Fig. 3.8.9), Assemble 2.1 finds 10 possible isomers (see Fig. 3.8.10):

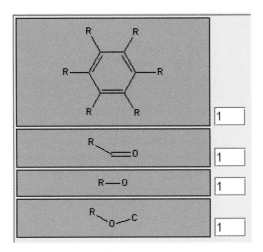

Fig. 3.8.9: Fragments entered to Assemble 2.1

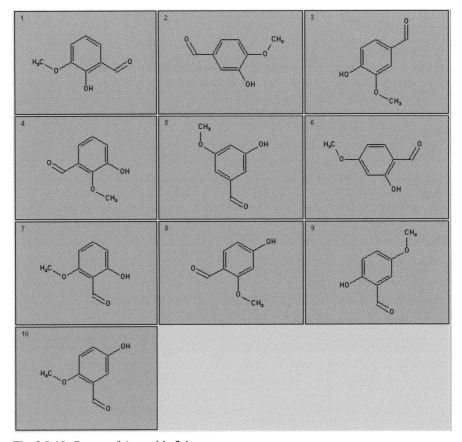

Fig. 3.8.10: Output of Assemble 2.1

The benzene ring is 1,2,3-trisubstituted in structures 1, 4, and 7 of Fig. 3.8.10, 1,3,5-trisubstituted in 5, and 1,2,4-trisubstitued in the remaining six structures. In the case of a first-order spectrum, the proton multiplicity would clearly differentiate the three types of substitutions from each other. The fact that one proton shows only one single large (*ortho*) and no *meta* coupling indicates a 1,2,4-substitution. However, the other two protons have very similar shifts so that we deal with a higher-order spectrum which can obscure their splitting, even though they are further away in the spectrum. The large coupling positively excludes only the 1,2,3-trisubstituted structures 1, 4, and 7.

The structure assignment of both isomers, A and B, is straightforward on the basis of their 1D NOESY spectra. Upon irradiating the CH_3O and CH=O protons, an increase in the intensity is expected for the aromatic H atoms that relax via dipolar interaction with the irradiated protons, i.e., for those in steric proximity. For isomer A, saturation of the CH_3O protons induces an increase in the intensity of the proton at δ 7.41 (Fig. 3.8.3, top). In a second experiment (Fig. 3.8.3, middle), the irradiation of the aldehyde H causes an increase in the intensity of the H atoms at δ 7.41 and 7.42. Therefore, the proton at δ 7.41 must be located between these two substituents (possible for structures 3, 5, and 9) and the aldehyde must have two protons in *ortho* position (as in 2, 3, and 5). Since 1,2,3-trisubstitution has been excluded because of the large coupling of the H atom at δ 6.97, structure 3 (vanillin) must be the correct solution.

The corresponding 1D NOESY spectra of isomer B (Fig 3.8.5, top and middle) show that the CH_3O group has one neighboring H at δ 6.97 and the aldehyde is in *ortho* position relative to the other two protons at δ 7.44 and 7.43. This is only possible for structure 2 (isovanillin).

3.8.3   Comments

3.8.3.1   Mass Spectra

For aldehydes, the loss of a H radical from the molecule typically is a favorable process, which is usually followed by the cleavage of the C=O group resulting in a less abundant fragment at $[M - 1 - 28]^+$ (here, m/z 123). For anisoles, characteristic processes are the loss of a methyl radical (to $[M - 15]^+$, here, m/z 137) followed by the elimination of C=O (to $[M - 15 - 28]^+ = [M - 43]^+$, here, m/z 109). The fragment at m/z 81 is probably due to a subsequent C=O elimination from the aromatic C–OH. Note that, as a general rule, the fragmentation of the C–X bond usually does not occur in aromatic alcohols, ethers, and amines. In these cases, the hetero atom is rather eliminated together with the aromatic carbon atom as C=O, HCO, or HCN.

3.8.3.2   Infrared Spectra

Generally, the O–H stretching vibration gives rise to a sharp band in the range of 3650–3590 cm^{-1} for the free form and has lower frequencies (3550–3450 cm^{-1}) for hydrogen-bonded species. The latter usually show broad bands because numerous different species are present. However, if one single hydrogen-bonded species occurs, as in the present cases, a sharp band can be observed at lower frequencies than expected for the free form (here, 3530/3545 cm^{-1}).

3.8.3.3 ^{1}H and ^{13}C NMR Spectra

Because of similar influences of the –OH and –OCH$_3$ substituents on the chemical shifts, it is not possible to derive the correct structures from the signal positions. Also, the HMBC spectrum does not provide straightforward information since some of the couplings over two and three bonds are seen but some others not (see Section 3.8.3.4). In such a case, the use of a 1D or a 2D NOESY spectrum is recommended. The signal intensities in the 1D NOESY spectra and the presence of 2D NOESY cross peaks can be used in a qualitative way indicating a distance of < ca. 5 Å between the two protons involved. While all relations between the H atoms are evident from a single 2D NOESY experiment, it has the drawback that, as usual in two-dimensional experiments, the digital resolution in the horizontal (F$_1$) dimension is limited. In the present case, the resolution of the NOESY spectrum (Fig. 3.8.11) is not sufficient for distinguishing between the two protons near δ 7.4.

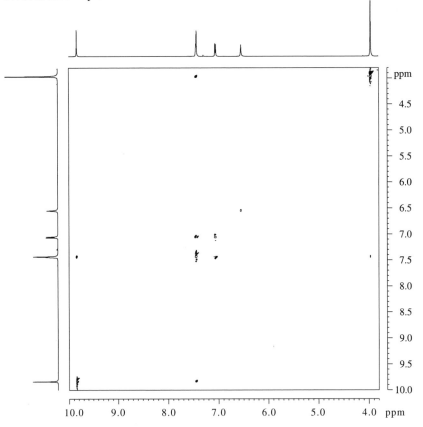

Fig. 3.8.11: NOESY spectrum of isomer A, 500 MHz, solvent CDCl$_3$

On the basis of the HMBC responses due to ^{13}C,^{1}H couplings over two and three bonds, we can fully assign ^{13}C NMR spectra. For isomer A (Fig. 3.8.7, bottom), a differentiation of the two deshielded quaternary aromatic atoms (δ$_{C-4}$ 151.9 and δ$_{C-3}$

147.2) is evident on the basis of the 3.94/147.2 cross peak. The doublet at δ_{H-5} 7.03 points over three-bond correlations to the signals of C-3 (δ 147.2) and C-1 (δ 129.2). Furthermore, due to the $^2J_{H-5,C-4}$ coupling, it also has a weak cross peak with the C-4 signal. From the expanded section of the HMBC spectrum, it is obvious that H-2 (δ 7.41) correlates with C-3, C-4, and C-6 and that H-6 (δ 7.42) has no correlation with C-3. For isomer B (Fig. 3.8.8, bottom), the differentiation of the two deshielded quaternary aromatic carbon atoms (δ_{C-4} 152.0 and δ_{C-3} 146.2) is evident on the basis of the 3.98/152.0 cross peak. The doublet at δ_{H-5} 6.97 again points over three-bond correlations to C-3 (δ 146.2) and C-1 (δ 130.6). From the expanded section of the HMBC spectrum, it is evident that H-2 (δ 7.44) is correlated with C-3, C-4, and C-6 and, on the other hand, H-6 (δ 7.43) with C-2, C-4, and C=O.

3.8.3.4   Presentation of NMR Data (500 resp. 125 MHz, CDCl$_3$, δ)

Isomer A

Assignment	^{1}H (J)	^{13}C	Selected HMBC responses (^{13}C partners)
1	–	129.2	–
2	7.41, d (ca. 2 Hz)	109.0	C-3, C-4, C-6, C=O
3	–	147.2	–
4	–	151.9	–
5	7.03, d (8.5 Hz)	114.4	C-1, C-3, C-4
6	7.42, dd (ca. 8, 2 Hz)	127.2	C-2, C-4, C=O
HC=O	9.81, s	191.1	C-1, C-2
CH$_3$O	3.94	55.7	C-3
OH	6.48, s	–	–

Isomer B

Assignment	^1H (J)	^{13}C	Selected HMBC responses (^{13}C partners)
1	–	130.6	–
2	7.44, d (ca. 2 Hz)	110.3	C-3, C-4, C-6
3	–	146.2	–
4	–	152.0	–
5	6.97, d (8.0 Hz)	114.1	C-1, C-3
6	7.43, dd (ca. 8, 2 Hz)	124.7	C-2, C-4, C=O
HC=O	9.83, s	191.2	C-1, C-2
CH$_3$O	3.98	56.2	C-4
OH	6.01, s	–	–

3.9 Problem 9

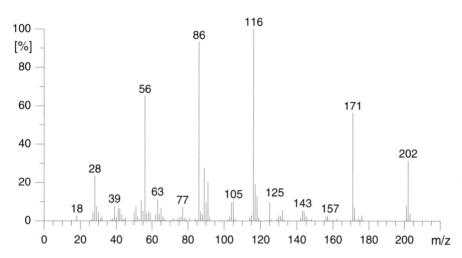

Fig. 3.9.1: Mass spectrum, EI, 70 eV

Fig. 3.9.2: IR spectrum, solvent CHCl$_3$, cell thickness 0.2 mm

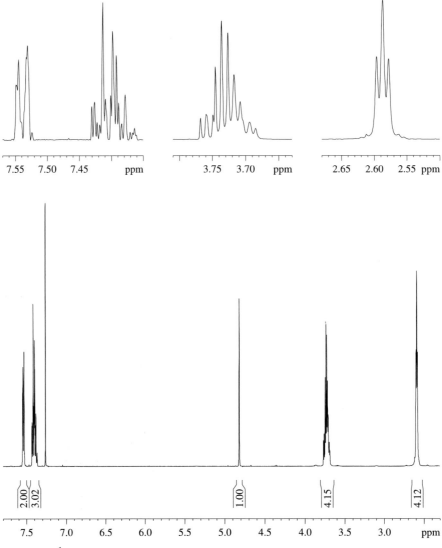

Fig. 3.9.3: ^{1}H NMR spectrum, 500 MHz, solvent CDCl$_3$

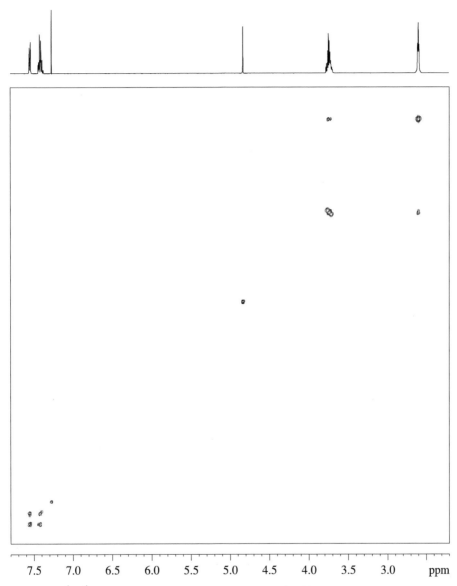

Fig. 3.9.4: ^{1}H,^{1}H COSY spectrum, 500 MHz, solvent CDCl$_3$

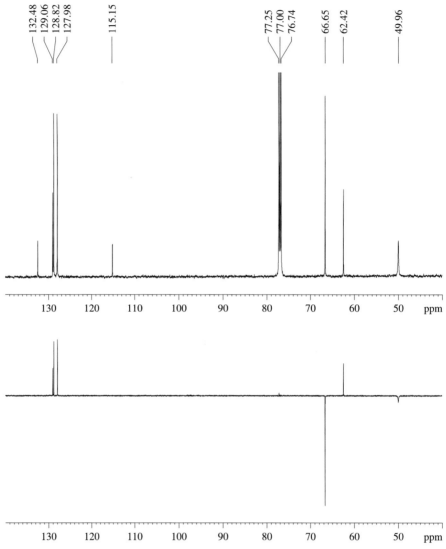

Fig. 3.9.5: ^{13}C NMR spectra, 125 MHz, solvent CDCl$_3$. Top: proton-decoupled; bottom: DEPT135

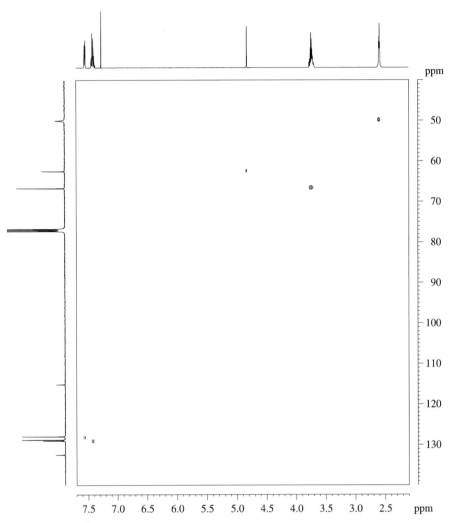

Fig. 3.9.6: $^{13}C,^{1}H$ HSQC spectrum, solvent CDCl$_3$

3.9.1   Elemental Composition and Structural Features

In the mass spectrum (Fig. 3.9.1), m/z 202 is assumed to be the molecular ion because all signals of lower mass can be related to it by chemically reasonable fragmentation. A ^{13}C isotope peak of 13% intensity relative to M$^{+\cdot}$ at m/z 203 limits the possible number of carbon atoms to twelve. The mass difference to the first fragment ion at m/z 171 indicates the presence of an sp^3-hybridized oxygen atom. The fragment series m/z 39, 51, 63 to 65, 77, 91, 104, 105 and maxima at m/z 28, 56, 86 suggest a mixed aromatic/nonaromatic nature of the compound.

The C–H stretching vibrations above and below 3000 cm^{-1} and the aromatic skeletal vibrations at 1500 and 1600 cm^{-1} in the IR spectrum (Fig. 3.9.2) corroborate the conclusions drawn from the mass spectrum. No OH, NH, or carbonyl groups are present.

The ^{1}H NMR spectrum (Fig. 3.9.3) shows 14 protons with integral ratios of 2:3:1:4:4 (from left). The signals at δ 7. 54 (dm, 2H) and δ 7.44–7.35 (m, 3H) may be assigned to a monosubstituted benzene ring, the singlet at δ 4.82 to a methine group attached to a hetero atom. In addition, there are two multiplets at δ 3.73 (4H) and 2.59 (4H). The COSY spectrum (Fig. 3.9.4) proves that the corresponding protons are coupled with each other. The five aromatic protons form a separate spin system. The signal at δ 7.54 with doublet-multiplet splitting can be assigned to the *ortho* protons.

In the proton-decoupled ^{13}C NMR spectrum (Fig. 3.9.5), four signals in the range of δ 133 to 115 (three =CH and one C) can be assigned to a monosubstituted benzene ring. There is one additional signal (at δ 132.5 or 115.2) in this region. We further find a signal for a CH at δ 62.4 and two for CH_2 groups at δ 66.7 and 50.0. According to the integration in the ^{1}H NMR spectrum, each of the latter ones corresponds to two methylene groups. The related ^{1}H and ^{13}C chemical shifts can be unambiguously assigned from the HSQC spectrum (Fig. 3.9.6). The chemical shifts δ 66.7/3.73 indicate that one methylene group is attached to an oxygen atom, whereas the other (δ 50.0/2.59) cannot be bonded to oxygen

So far, we have found $C_{12}H_{14}O$ as partial formula, corresponding to a mass of 174 u. The missing 28 u can be accounted for by two nitrogen atoms or by an additional C and O, leading to two possible molecular formulas, namely $C_{12}H_{14}N_2O$ or $C_{13}H_{14}O_2$, either of them with seven double bond equivalents. No convincing indications as to which of the two is the correct one are readily available at this stage.

3.9.2   Structure Assembly

First, we consider all possible isomers having the molecular formula $C_{13}H_{14}O_2$, a monosubstituted benzene ring and two $CH_2CH_2(O)$ groups as substructures, a symmetry requiring eight signals in the ^{13}C NMR spectrum, and no C=O group.

Check with Assemble 2.1
Start Assemble 2.1 and enter the molecular formula and the above-listed structural information. Use the neighbor atom tag to describe the presence of an oxygen atom next to one of the CH_2 groups. If this is not explicitly stated, the required twofold occurrence of the substructure CH_2CH_2O would attribute two different oxygen atoms to the two CH_2CH_2 groups and exclude solutions having a $CH_2CH_2OCH_2CH_2$ group. The structures obtained are shown in Fig. 3.9.7. Since Assemble 2.1 only generates topological isomers, compound 5 implies two different structures.

Use the ^{13}C NMR ranking module to automatically compare the estimated and observed chemical shifts. The best solution has mean and maximum deviations of 7.20 and 24.3 ppm, respectively.

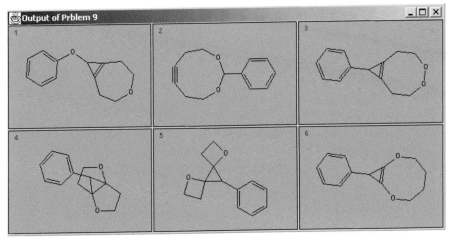

Fig. 3.9.7: Output of Assemble 2.1 with the molecular formula $C_{13}H_{14}O_2$

Repeat the isomer generation with the molecular formula $C_{12}H_{14}N_2O$ and the same structure information as above. Ranking of the 18 isomers thus generated will show one compatible solution with mean and maximum deviations of 0.5 and 1.9 ppm, respectively, of the ^{13}C NMR shifts. The second best solution already has mean and maximum deviations of 4.50 and 31.6 ppm, respectively. The correct solution, therefore, is:

3.9.3   Comments

3.9.3.1   Mass Spectrum

In principle, the intensity of the ^{13}C isotope peak could be used to infer the number of carbon atoms in the molecule since each carbon atom contributes 1.1% to the first isotope peak. However, this isotope peak may gain intensity owing to the protonation of the molecular ion, thus simulating too many carbons. Therefore, only the upper limit of the number of carbon atoms can be inferred. Furthermore, intensity values in mass spectra are of limited precision due to various instrumental and operating artifacts. Thus, valid conclusions can only be drawn if the deviation between found and expected intensity of $[M + 1]^+$ is large. In the present case, an intensity of 12.6% formally limits the maximum number of carbon atoms to 11, but both molecular formulas are within the tolerance.

The mass spectrum can be rationalized as showing mainly the peaks corresponding to the two moieties of the molecule obtained by benzylic cleavage (morpholine ring: m/z 86, aromatic moiety: m/z 116) and by loss of formaldehyde ($\Delta m = 30$) from $[M - 1]^+$ and from m/z 86 in a retro-Diels–Alder-type reaction as follows:

$$R: \quad -H \quad (m/z\ 56)$$

Loss of the phenyl group from the molecular ion yields the peak at m/z 125.

3.9.3.2 Infrared Spectrum

A nitrile group may be expected to give rise to a sharp signal in the range of 2260–2230 cm^{-1} and is, in general, easily identified. In fact, there is a signal at 2230 cm^{-1} but at first glance, it is too weak in this case to be recognized as indicating nitrile, especially because overtone and combination bands of similar intensity are present in its neighborhood and because it might as well be interpreted as an overtone of the strong band at 1120 cm^{-1}. It is generally observed that nitrile absorptions are scarcely visible if the group is attached to a carbon atom carrying an electronegative substituent.

The various bands of low intensity between 2500 and 1500 cm^{-1} may be rationalized as follows: The broad band centered at 2450 cm^{-1} is an instrumental artifact due to solvent absorption at this frequency. The band at 2230 cm^{-1} is assigned to the CN stretching vibration of the nitrile group. The double band near 1970 cm^{-1} and the band at 1810 cm^{-1} are due to various combination vibrations and overtones of the benzene moiety. The band at 1600 cm^{-1}, finally, is a skeletal vibration of the benzene ring. The very strong and comparatively sharp band at 1120 cm^{-1} is due to a vibration of the morpholino part of the molecule, which is probably best described as C–O–C asymmetric stretching vibration.

3.9.3.3 ^{1}H NMR Spectrum

Since the molecule is chiral, the geminal methylene protons are, in principle, anisochronous even if there is a fast ring inversion on the corresponding NMR time scale. A fast rotation around the CH–N bond together with a fast ring inversion cause an equivalence of each of the protons of one pair of methylene groups with the counterpart on the other pair of methylene groups (see Chapter 4.3). Thus, in the general case, if fast rotation and ring inversion occur, an $AA'BB'CC'DD'$ spin system is to be expected. This is simplified to two identical $ABCD$ spin systems if all couplings across the hetero atoms disappear. The actual spectrum clearly shows the nonequivalence of the low-field protons, while the apparent triplet of the high-field methylene groups is a consequence of an accidental equivalence. A careful investigation of this signal indicates that it must have a higher multiplicity, but the lines are not resolved. Also, the small signals on both sides of the triplet are caused by higher-order effects.

3.9.3.4 ^{13}C NMR Spectrum

The chemical shift at δ 115.2 provides strong evidence for the presence of a cyano group because no alternative assignment is possible under the given premises. Only an additional aromatic or heteroaromatic ring, a symmetrically persubstituted carbon–carbon double bond, or a carbon atom between two oxygen or halogen atoms could, otherwise, account for this chemical shift value.

There are 8 signals in the ^{13}C NMR spectrum because two pairs of aromatic CH and of morpholino CH_2 are isochronous. Although the molecule is chiral, a fast rotation of the phenyl and morpholino groups causes chemical shift equivalence of the permutated atoms and one would naively expect double intensities. However, differences in relaxation times and the nuclear Overhauser effect often bias line intensities (see Chapter 4.3). Nevertheless, identification of the *para* carbon atom of the benzene ring is obvious. For the morpholino group, the situation is more complicated because the axis of rotation becomes a symmetry element only because of fast conformation changes in the 6-membered ring. One of the processes is slow enough to cause line broadening of the signal at δ 50.0 and, therefore, a reduced signal height. The integrated intensity, however, would probably be as expected, i.e., the same as for the other CH_2 signal. Sharper lines would be observed at higher temperatures.

3.9.3.5 Presentation of NMR Data (500 resp. 125 MHz, CDCl$_3$, δ)

Assignment	^{1}H	^{13}C
CH	4.82, s, 1H	62.4
NCH$_2$	2.59, m, 4H	50.0
OCH$_2$	3.73, m, 4H	66.7
C$_{ipso}$	–	132.5
C$_{ortho}$	7.54, dm, 2H	128.0
C$_{meta}$	7.41, m, 2H	128.8
C$_{para}$	7.37, m, 1H	129.1
CN	–	115.2

3.10 Problem 10

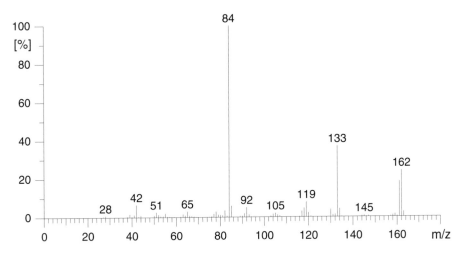

Fig. 3.10.1: Mass spectrum, EI, 70 eV

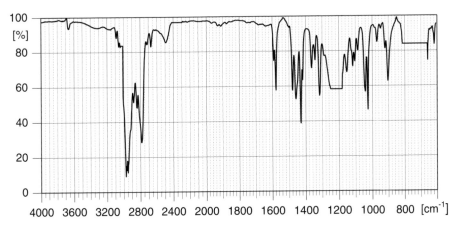

Fig. 3.10.2: IR spectrum, solvent CHCl₃, cell thickness 0.2 mm

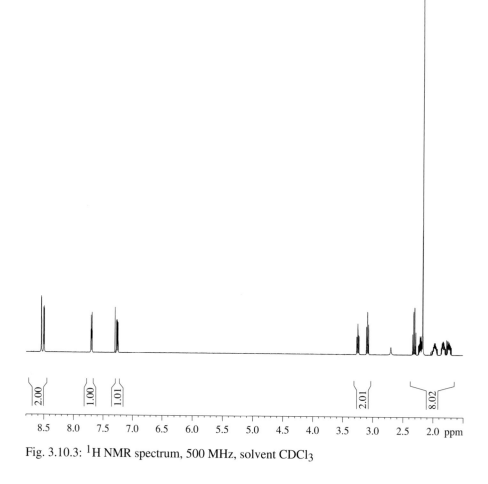

Fig. 3.10.3: ^{1}H NMR spectrum, 500 MHz, solvent CDCl$_3$

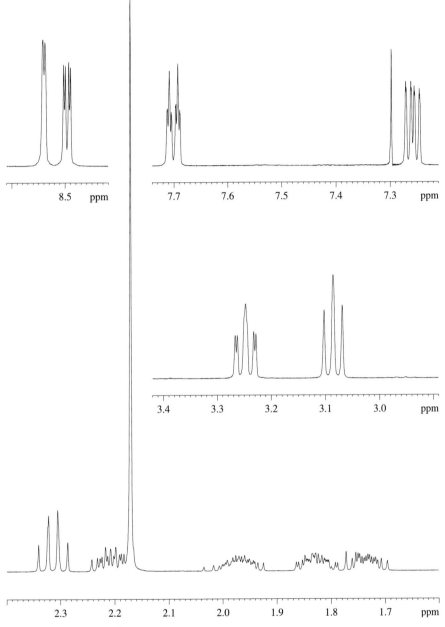

Fig. 3.10.4: Expanded ^{1}H NMR spectrum, 500 MHz, solvent CDCl$_3$

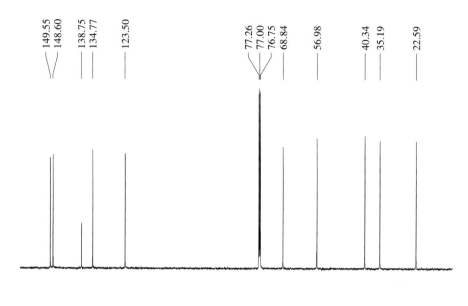

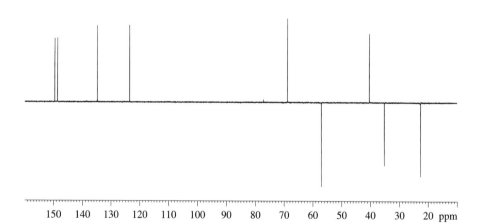

Fig. 3.10.5: ^{13}C NMR spectra, 125 MHz, solvent CDCl$_3$. Top: proton-decoupled; bottom: DEPT135

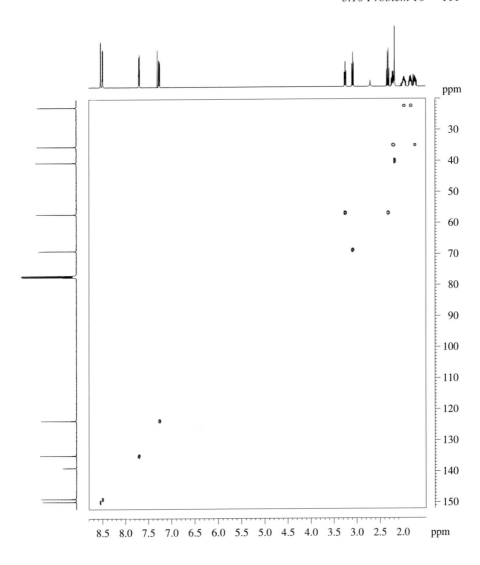

Fig. 3.10.6: $^{13}C,^{1}H$ HSQC spectrum, solvent CDCl$_3$

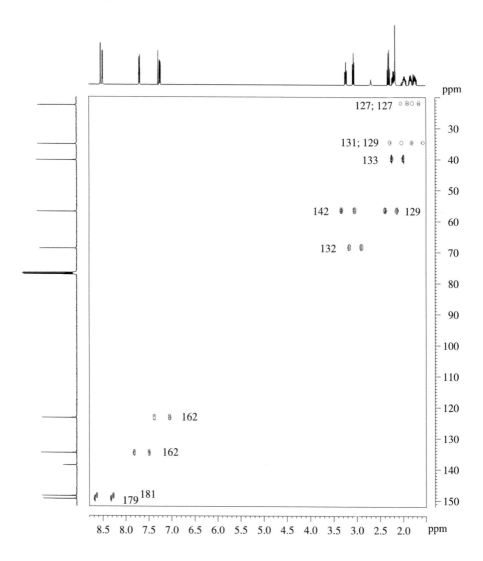

Fig. 3.10.7: $^{13}C,^{1}H$ HSQC spectrum with $^{1}J_{C,H}$, solvent CDCl$_3$

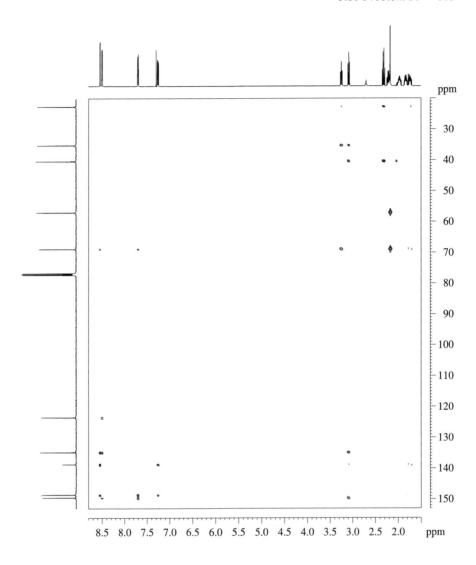

Fig. 3.10.8: $^{13}C,^{1}H$ HMBC spectrum, solvent CDCl$_3$

3.10.1 Elemental Composition and Structural Features

In the IR spectrum (Fig. 3.10.2), we note two uncommon bands at 2670 and 2500 cm^{-1}. The signal at 1590 cm^{-1} could indicate an aromatic moiety. Note that most of the typical bands in the regions of 1250–1180 and 820–660 cm^{-1} are hidden by solvent absorption.

From the mass spectrum (Fig. 3.10.1), we assume a molecular mass of 162 u, the most abundant peaks at m/z 133 and 84 corresponding to a loss of 29 and 78 u, respectively. The even mass number of the base peak means that it is either produced through a rearrangement or that it contains an odd number of nitrogen atoms. In the latter case, the molecule has to contain an even number of N atoms to account for the even molecular mass.

Ten signals are discernible in the ^{13}C NMR spectrum (Fig. 3.10.5, top), indicating the presence of at least ten carbon atoms (if no signal splitting occurs due to coupling with nuclei other than protons). Five signals fall in the range expected for aromatic carbons (δ ca. 160–100), two of them exhibiting rather extreme chemical shift values close to δ 150. We may, thus, tentatively assume that the corresponding carbon atoms are substituted with hetero atoms. Since they carry a hydrogen atom, as shown by the DEPT135 (Fig. 3.10.5, bottom) and HSQC spectra (Fig. 3.10.6), the hetero atom is part of the aromatic moiety. These spectra indicate five methine groups (four aromatic, one aliphatic), three methylene groups, one methyl group, and one quaternary (aromatic) carbon atom. We, therefore, expect at least 14 H atoms in the ^{1}H NMR spectrum. The HSQC spectrum directly gives the ^{1}H and ^{13}C chemical shifts of the corresponding CH units. From the HSQC spectrum without ^{13}C decoupling (Fig. 3.10.7), we obtain the ^{13}C,^{1}H cross peaks with a splitting due to ^{13}C-^{1}H coupling over one bond. The size of these $^{1}J_{C,H}$ coupling constants depends on the hybridization of the carbon atom (for benzene, it is 159 Hz). Furthermore, their values are strongly increased by substitution with electronegative substituents. The values of $^{1}J_{C,H}$ = 181 and 179 Hz for the =CH moieties with ^{13}C chemical shifts close to δ 150 also corroborate that they are substituted with a hetero atom.

Integration of the signals in the ^{1}H NMR spectrum (Fig. 3.10.3) gives a hydrogen ratio of 1:1:1:1 in the aromatic region (δ 8.6 to 7.2) and 1:1:1:4:1:2 (from left) in the aliphatic region, summing up to 14 H atoms. This is consistent with the information from the ^{13}C NMR spectrum and indicates that all protons are attached to carbon atoms. The assignment of the sharp singlet at δ 2.17 with an intensity of 3H (from the HSQC spectrum, Fig. 3.10.6: δ_C 40.3) to a methyl group is straightforward.

Summing up the elemental composition given by the partial structures identified so far, yields $C_{10}H_{14}$ plus, at least, one hetero atom. Thus, the difference to the molecular mass amounts to 162 – 134 = 28 u including at least one hetero atom. Since C_2H_4 can be excluded because it lacks a hetero atom and CO would demand an additional line in the ^{13}C NMR spectrum, N_2 is the only remaining choice for assigning this difference. Hence, the elemental composition of the compound is $C_{10}H_{14}N_2$.

3.10.2 Structure Assembly

We now proceed to assemble a tentative constitution from the fragments found. We expect a heteroaromatic system with five carbon atoms. In the actual context, this calls for a pyridine ring. As four aromatic protons are present, it must be monosubstituted. This is corroborated by the prominent loss of 78 (corresponding to C_5H_4N) from the molecular ion in the mass spectrum. Benzene can be ruled out as a possible assignment for Δm = 78 because the ^{13}C NMR spectrum is not compatible with a phenyl ring. The other fragment at m/z 84 has an elemental composition of $C_5H_{10}N$ and contains one double bond equivalent and one methyl group. The methyl group, giving a singlet in the ^{1}H NMR spectrum at δ 2.17, has to be bonded to the nitrogen atom. As no other terminal groups are available, the nitrogen atom must be part of the ring. Furthermore, chemical shifts of δ 68.8 and 57.0 in the ^{13}C NMR spectrum for methine and methylene groups, respectively, indicate the following partial structure:

$$CH_3-N \overset{\displaystyle CH-}{\underset{\displaystyle CH_2-}{\big<}} \quad \Big\} \quad \text{ring}$$

Check with Assemble 2.1

Start Assemble 2.1, enter the molecular formula $C_{10}H_{14}N_2$, the fragments for a monosubstituted pyridine and $CH_3N(CH_2R)CHR_2$, and the atom constraints (see Fig. 3.10.9). Since the position of the substituent is not defined, free valences (represented by R) must be entered at three positions of the pyridine fragment. The atom tag of the CH stating that only one hydrogen atom is attached to this carbon atom is required since otherwise one of the substituents R could be replaced by H and two more structures would be generated.

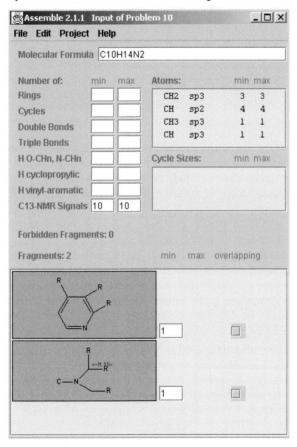

Fig. 3.10.9: Input for Assemble 2.1

The six solutions found are shown in Fig. 3.10.10. For symmetry reasons, no 4-substituted pyridine is generated.

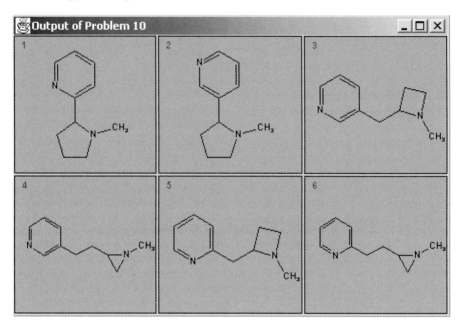

Fig. 3.10.10: Output of Assemble 2.1

A clear distinction among the structures can be achieved on the basis of the HMBC spectrum (Fig. 3.10.8). The correlations of the methine proton *via* two and three bonds to three aromatic carbon atoms exclude the structures 3–6. Similar conclusions can be drawn from the mass spectrum, where a prominent fragmentation would take place between the two methylene groups in structures 4 and 6 and between the methylene group and the 4-membered ring in structures 3 and 5 (benzylic cleavages), both leading to m/z 60 and/or 92 instead of m/z 84 and 78.

From the two remaining isomers 1 and 2, the HMBC correlations of the methine proton with three aromatic signals at δ 138.8 (quaternary), 134.8, and 149.6 unambiguously prove the substitution in position 3. The last signal with its extreme deshielding must correspond to the carbon attached to nitrogen. In case of structure 1, only two cross peaks would occur, whereas for substitution at position 4 (which was excluded for symmetry reasons), the HMBC cross peak with the NCH carbon atom is not possible. The unknown compound at hand is, therefore, nicotine:

Check with Assemble 2.1

Use the Rank Output command of the Project menu and enter the required ^{13}C NMR information. The correct solution is found in the first position, with 0.8 and 2.1 ppm as mean and maximum deviations, respectively. For ^{1}H NMR, use the average shifts of the geminal protons since the ranking module only makes use of the topology and predicts identical chemical shifts for diastereotopic protons. Here too, the correct structure is found as the best one, with mean and maximum deviations of 0.22 and 0.50 ppm, respectively.

3.10.3 Comments

3.10.3.1 Mass Spectrum

The statement that an even-mass fragment originating from an even-mass molecular ion must be formed by H rearrangement or otherwise contain an odd number of nitrogen atoms is slightly oversimplified, but implies a general rule: Fragments formed by a single bond cleavage in a radical cation are of even mass only if they contain an odd number of atoms whose atomic weight and number of valencies are not both either even or odd. Since nitrogen is the only element in common organic compounds with an even atomic weight (14) and an odd number of valencies (3), the rule is usually applied as stated above. It assumes that no heavy isotopes as, e.g., ^{13}C, ^{15}N, [17]O, [2]H (deuterium) etc. and no unusual elements such as Fe(III) and the like are present. Reactions involving multiple bond cleavages do, of course, require adaptation of the rule, e.g., *retro*-Diels–Alder reactions (equivalent to two single bond cleavages) yield even-mass fragments without nitrogen being present, and double H rearrangements, accordingly, give odd-mass products from even-mass educts.

Loss of ethyl radical from nicotine (formation of the prominent peak at m/z 133) is a common reaction in saturated alicyclic amines. The same holds true for the loss of hydrogen radical to yield $[M - H]^+$.

3.10.3.2 Infrared Spectrum

Methyl groups and often also methylene groups bonded to a nitrogen atom in an amine exhibit characteristically low C–H stretching frequencies at or even below 2850 cm^{-1}. The present spectrum shows this very nicely. There are, however, other structural elements that also exhibit C–H stretching frequencies below 2850 cm^{-1}.

The band at 2500 cm^{-1} can be assigned to the NH^+ stretching vibration of the protonated nicotine, which is formed either with traces of HCl from the solvent chloroform or with CO_2 and humidity from the air. Some protonated derivatives are usually found in the spectra of aliphatic and alicyclic amines. The very weak band at 3650 cm^{-1} together with the bumpy baseline from 3400–3200 cm^{-1} indicate traces of water.

3.10.3.3 ^{1}H and ^{13}C NMR Spectra

An unambiguous assignment of the ^{1}H and ^{13}C signals can be obtained from the HMBC spectrum in combination with the results derived from the HSQC measurements

(Figs. 3.10.6 and 3.10.7). We only need a safe starting point, e.g., the CH_3 protons. The cross peaks of the methine (δ 68.8) and methylene (δ 57.0) carbon signals give the assignment of C-2' and C-5'. The HMBC response of H-2' (δ 3.09) with the only quaternary aromatic carbon atom proves that the chemical shift of C-3 is δ 138.8. The cross peaks at δ 3.09/134.8 and δ 3.09/149.6 give the assignments of C-4 and C-2, respectively. The other signal with extreme ^{13}C chemical shift (δ 148.6) must, therefore, belong to C-6, and the remaining line in the aromatic region (δ 123.5) corresponds to C-5. A differentiation between the C-3' and C-4' methylene groups follows from the 1.73/138.8 cross peak because only the H-3' methylene protons can give an HMBC correlation over three bonds with the C-3 carbon. Taking into account the deshielding effect of the aromatic substituent in β-position, the higher chemical shift for C-3' is also obvious. Usually, C-H coupling constants over 4 bonds are < 3 Hz and do not give rise to corresponding correlations in HMBC spectra obtained with the usual settings of the experimental parameters. However, as in the case of H-H couplings, the C-H coupling constants over four σ bonds are larger if the bonds of the coupling path are arranged in a W form. The cross peak of one of the C-4' protons with the methyl group (δ 1.97/40.3) corresponds to such a coupling.

In pyridine derivatives, there is a significant difference between the vicinal coupling constants, $^3J_{\text{H-2,H-3}}$ = 4.8 Hz and $^3J_{\text{H-3,H-4}}$ = 7.9 Hz. Due to the electronegative nitrogen atom attached to C-2, the first coupling is strongly reduced, whereas the second one is slightly increased relative to the corresponding value of benzene. The strong coupling of the protons of the pyrrolidine moiety leads to a rather complex spectrum, in which the signal pattern is further complicated by the fact that the molecule is chiral and, thus, the methylene protons are diastereotopic.

Check with Assemble 2.1
By pressing the right mouse button over a structure, a menu pops up from which you can select "Calculate 3D Coordinates". The resulting structure shown in Fig. 3.10.11 explains why the chemical shift values of the two protons of CH_2N are so different: One of the protons is close to the aromatic ring, whereas the other is far away.

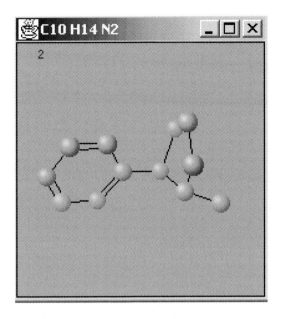

Fig. 3.10.11: Three-dimensional model of nicotine generated by Assemble 2.1

3.10.3.5   Presentation of NMR Data (500 resp. 125 MHz, CDCl$_3$, δ)

Assignment	^{1}H (J)	^{13}C ($^1J_{C,H}$)	HMBC responses (^{13}C partners)
2	8.54, d (2.0 Hz)	149.6 (179)	C-3, C-4, C-6, C-2'
3	–	138.8	–
4	7.70, dt (7.9, 2.0 Hz)	134.8 (162)	C-2, C-6, C-2'
5	7.26, dd (7.9, 4.8 Hz)	123.5 (162)	C-3, C-6
6	8.50, dd (4.8, 2.0 Hz)	148.6 (181)	C-2, C-4, C-5
2'	3.09, t (8.4 Hz)	68.8 (132)	C-2, C-4, C-3', NCH$_3$
3'	2.21, m	35.2 (131)	–
	1.73, m	(129)	C-3, C-2', C-4'
4'	1.97, m	22.6 (127)	NCH$_3$
	1.83, m	(127)	–
5'	3.25, ddd (9.5, 8.4, 2.0 Hz)	57.0 (142)	C-2', C-3', C-4'
	2.31, dt (9.5, 8.4 Hz)	(129)	C-4', NCH$_3$
NCH$_3$	2.17, s	40.3 (133)	C-2', C-5'

3.11 Problem 11

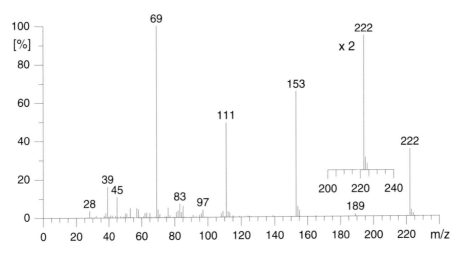

Fig. 3.11.1: Mass spectrum, EI, 70 eV

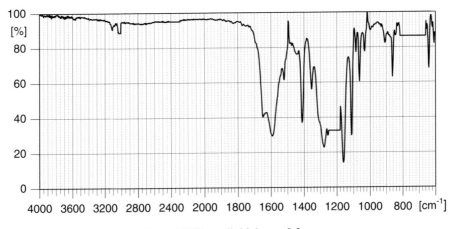

Fig. 3.11.2: IR spectrum, solvent CHCl₃, cell thickness 0.2 mm

UV spectrum (solvent ethanol):

	λ_{max}	$\log \varepsilon$
	265	3.96
	290	3.88
	332	3.63

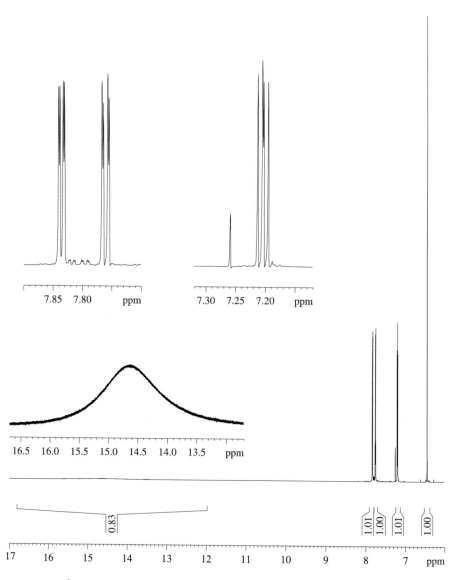

Fig. 3.11.3: ^{1}H NMR spectrum, 500 MHz, solvent CDCl$_3$

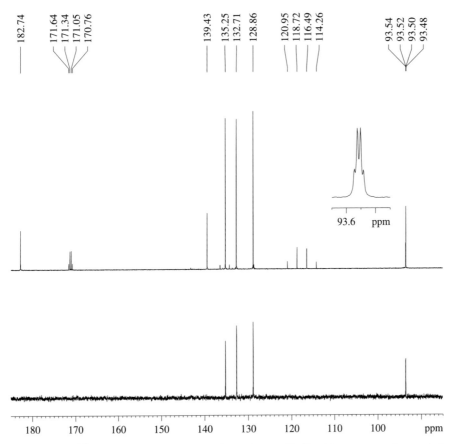

Fig. 3.11.4: ^{13}C NMR spectra, 125 MHz, solvent CDCl$_3$. Top: proton-decoupled; bottom: DEPT135

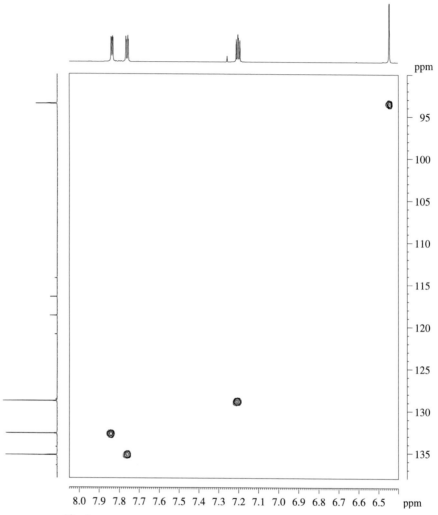

Fig. 3.11.5: $^{13}C,^{1}H$ HSQC spectrum, solvent CDCl$_3$

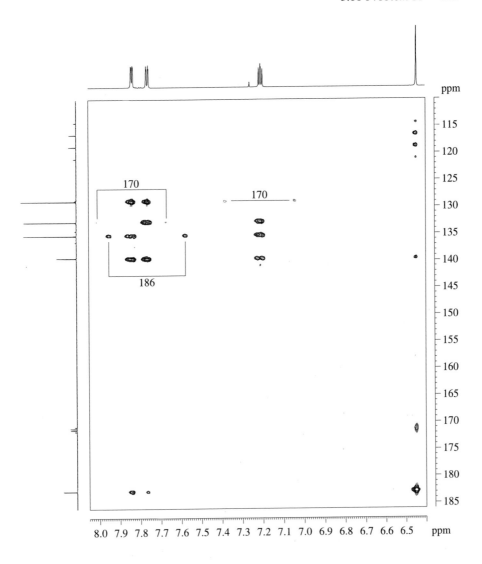

Fig. 3.11.6: ^{13}C,^{1}H HMBC spectrum, solvent CDCl$_3$

3.11.1 Elemental Composition and Structural Features

The integration of the ^{1}H NMR spectrum (Fig. 3.11.3) leads to intensity ratios of ca. 1:1:1:1:1. The broad signal at δ ca. 14.7 suggests the presence of a strongly hydrogen-bonded hydroxyl group. The three doublets of doublets between δ 7 and 8 represent an *AMX* spin system. There are no aliphatic protons in the molecule.

The ^{13}C NMR spectrum (Fig. 3.11.4, top) consists of 14 lines, four of which belong to CH groups according to the DEPT135 (Fig. 3.11.4, bottom) and HSQC (Fig. 3.11.5) spectra. The ^{1}H and ^{13}C chemical shifts of these =CH moieties correspond to δ 7.84/132.7, 7.76/135.3, 7.21/128.9, and 6.45/93.5. The presence of four methine groups as well as of one proton not bonded to a carbon atom is in full accordance with the ^{1}H NMR spectrum.

The IR spectrum (Fig. 3.11.2) shows a very broad absorption in the range of 3500–2000 cm^{-1}, confirming the existence of a strongly hydrogen-bonded hydroxyl group. Because of its extremely low intensity, it cannot originate from a carboxyl group. The strong bands at 1650–1590 cm^{-1} and 1300–1100 cm^{-1} are not easily assignable at this stage.

The mass spectrum (Fig. 3.11.1) ends with an intensive peak at m/z 222. There is no reason not to assume that this peak corresponds to the molecular ion. The intensity of the isotope signal at m/z 224 (ca. 5%) suggests that the molecule contains one sulfur atom. Silicon would also lead to a signal of similar intensity at [M + 2]$^+$ but at the same time it would also contribute with 5% to [M + 1]$^+$. The relative intensity of the latter signal is, however, only 10%, indicating that silicon is not present and that the molecule cannot contain more than nine carbon atoms, which is in conflict with the findings from the ^{13}C NMR spectrum having 14 lines. This too large number of signals could be explained by the presence of magnetic nuclei other than protons, which lead to splittings of the carbon signals due to spin coupling. One hint of the presence of such nuclei is given by the mass spectrum. The two most intense peaks are m/z 69 and m/z 153 (M$^+$ – 69). If there are no aliphatic protons in the molecule, as is the case here, these fragments are characteristic of a trifluoromethyl group.

The ^{13}C chemical shift of a trifluoromethyl group, generally, is in the range of δ 100–120 and the ^{13}C–^{19}F coupling constant is ca. 200–300 Hz. Since the group contains three magnetically equivalent coupling partners, we expect a quartet. Indeed, we find four lines (δ 121.0, 118.7, 116.5, 114.3) with a spacing of 2.23 ppm corresponding to 279 Hz in a 125 MHz spectrum. A closer inspection of the spectrum shows another quartet at ca. δ 171 which can be assigned to the carbon atom next to the trifluoromethyl group. According to the DEPT135 and HSQC spectra, it is a quaternary carbon atom. The 14 lines in the ^{13}C NMR spectrum, thus, correspond to only eight carbon atoms. The fragments found so far sum up to a molecular formula of $C_8H_5F_3OS$ accounting for a mass of 206 u.

3.11.2 Structure Assembly

The following structural elements have been detected:

Structure fragment	Mass
CF_3-C	81
CH–CH–CH	39
CH	13
OH	17
S	32
2C	24
Total mass	206

The difference of 16 u relative to the molecular mass of 222 u suggests the presence of a further oxygen atom. The molecular formula, thus, is $C_8H_5F_3O_2S$ and corresponds to five double bond equivalents (if a divalent sulfur atom is assumed).

One of the carbon atoms (CF_3) and one oxygen are sp^3-hybridized. With the remaining seven carbon atoms and one each of oxygen and sulfur, we may compose four double bonds at most. Since allenes and acetylenes can be excluded on the basis of the ^{13}C chemical shifts (both would have two lines below δ 90), the compound must have at least one ring which, moreover, has to contain the CH–CH–CH moiety. A benzene ring can be excluded since there are only four signals in the aromatic region. Considering the ^{13}C chemical shifts, ^1H-^1H coupling constants, and UV absorbances, the nonaromatic moieties cyclopentadiene and a 6-membered thiolactone can be excluded as well. Thus, the molecule must have a 5-membered heteroaromatic ring which may be either furan or thiophene.

Furan is less probable on the basis of the ^1H chemical shifts (a stronger shielding of all protons would be expected) and ^1H-^1H coupling constants (there is only one coupling constant larger than 3 Hz in furan, but here we observe two such couplings for the signal at δ 7.21) as well as from the ^{13}C chemical shifts (a stronger shielding occurs in furan). Also, the intensive fragment at m/z 111 in the mass spectrum (with a sulfur isotope signal at m/z 113) and the fragment at m/z 83 perfectly fit a thiophene derivative. Hence, no C=S group can be present so that the two signals of δ > 170 must belong to sp^2-hybridized carbon atoms bonded to oxygen. Since there are only two oxygen atoms and one of them is OH, we expect the molecule to contain one enol and one carbonyl group.

Check with Assemble 2.1

Start Assemble 2.1, enter the molecular formula of $C_8H_5F_3O_2S$ and the fragments identified above (see Fig. 3.11.7). The default valence of S is 2 but one can set a different value in the window popping up after starting the calculation (not shown), e.g., to generate sulfinyl or sulfonyl derivatives. If the absence of a substructure is entered as an input item, one must check the "Overlapping" box and set 0 both for the minimum and maximum numbers (second substructure in Fig. 3.11.7).

With this input, Assemble 2.1 generates nine structures (see Fig. 3.11.8; another set of 9 isomers is obtained if the thiophene ring in the third substructure is substituted at position 3). Structures 3 and 8 are tautomeric enol forms of a β-diketone. Also, isomers 1 and 7 as well as 5 and 6 are tautomeric pairs. They can be excluded together with structure 4 because their two deshielded carbon atoms (δ 182.7 and 171.2) are not attached to a hydrogen. Solutions 2 and 9 are enol forms of α-diketones, which would occur in their keto form.

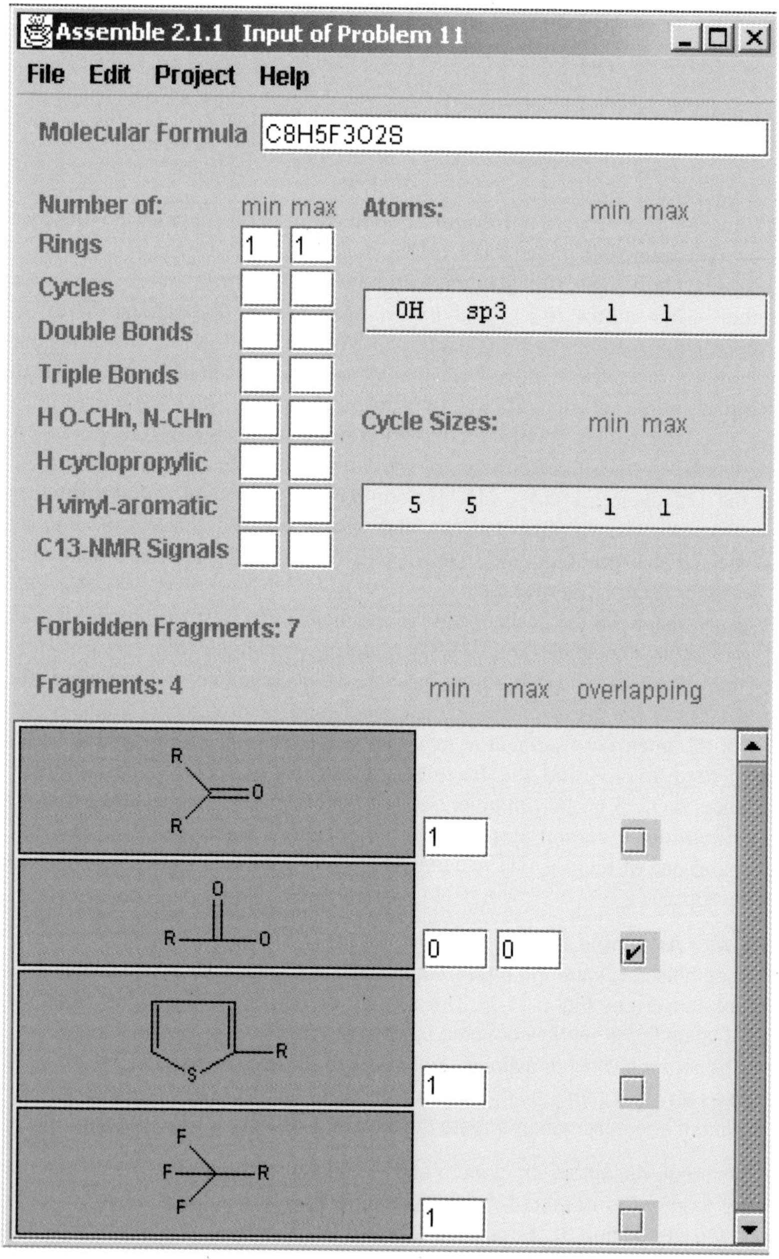

Fig. 3.11.7: Input to Assemble 2.1

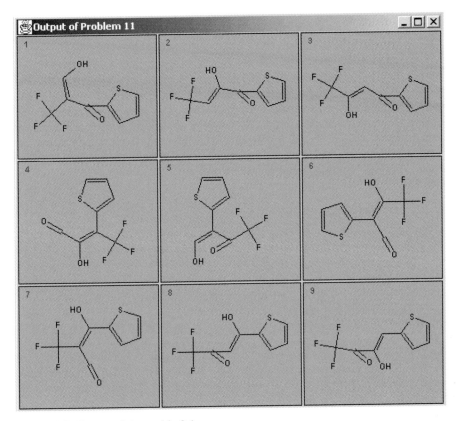

Fig. 3.11.8: Output of Assemble 2.1

According to the above discussion, there only remain structures 3 and 8, the tautomeric forms of 2-thenoyltrifluoroacetone:

I   II

The substituent must be attached in position 2 of the thiophene ring as indicated by the signal at δ 7.21 in the ^{1}H NMR spectrum, which shows two couplings above 3 Hz. For the 3-substituted thiophene, only one such coupling of the high-field proton would occur.

3.11.3 Comments

3.11.3.1 Mass Spectrum

The main features of the mass spectrum seem to arise by simple bond cleavages next to the carbonyl groups. The trifluoromethyl group is lost to give m/z 153 and appears as a charged fragment at m/z 69, which probably also results in part through decarbonylation of the acyl ion at m/z 97. The peak at m/z 111, which could be either half of the molecule, seems to be entirely due to the thiophenoyl ion as judged from the ^{34}S isotope peak at m/z 113. Both the appearance of m/z 189 (loss of SH from $M^{+\cdot}$) and formation of CHS^+ (m/z 45) could be taken as additional indications of the presence of sulfur. There is no sign of enolization in the gas phase.

3.11.3.2 Infrared Spectrum

Intermolecular hydrogen bonding of OH groups generally gives rise to a moderately sharp band between 3600 and 3450 cm^{-1}. If more than one hydrogen bond is involved with a particular hydroxyl group, the band becomes broader and moves to lower frequencies (3400–3200 cm^{-1}). Intermolecular hydrogen bonding is characterized by the band shape and frequencies being highly dependent on the sample concentration. Intramolecular hydrogen bonding, on the other hand, is not sensitive to sample concentration but it may be influenced by the solvent. In the general case, a band similar to that observed with intermolecular hydrogen bonding is found. However, if the hydrogen bonds are extremely strong as, e.g., in enolized β-diketones or in o-nitrophenols, very broad bands extending from 3200 down to 2000 cm^{-1} result. Since the total intensity is then spread over a wide range, such bands may easily be overlooked.

Carbonyl stretching frequencies are lowered by hydrogen bonding. If very strong intramolecular hydrogen bonds are formed as, e.g., in enolized β-diketones, the stretching frequencies of the two CO and two CC bonds in the enol molecule become similar and strong interactions occur. Therefore, it is no longer possible to assign the resulting broad absorption at about 1600 cm^{-1} to either the hydrogen-bonded carbonyl or the C=C double bond, but it must rather be attributed to the whole conjugated chromophore. Another group of strong bands arising predominantly from C–O stretching vibrations is observed around 1250 cm^{-1}.

The carbon-fluorine stretching frequencies fall into the range from 1400–1000 cm^{-1}. Due to rotational isomerism, several sharp bands, sometimes enfolded in a broad strong band are observed. As other chromophores give rise to intense absorptions in this region, fluorine is difficult to identify from the IR spectrum.

3.11.3.3 Ultraviolet Spectrum

The UV spectrum of the compound under study can hardly be rationalized without extensive calculations. However, it fits the general picture of thiophenes substituted with a carbonyl group in position 2. For such compounds, at least two bands are expected, separated by some 20 nm. Both bands have an extinction coefficient, ε, around 10 000, the lower wavelength band in general being slightly more intense.

3.11.3.4 ^{1}H and ^{13}C NMR spectra

An unambiguous assignment of the two deshielded thiophene protons is possible on the basis of the respective $^1J_{C,H}$ coupling constants, which can be read from the HMBC spectrum (Fig. 3.11.6) as 170 Hz for the signal at δ 7.84 (position 3) and 186 Hz for that at δ 7.76 (position 5). Because of its chemical shift of δ 7.21, the third proton must be at position 4 ($J_{C,H}$ = 170 Hz).

The cross peak at δ 7.76/182.7 corresponds to a coupling over four bonds. Such couplings occasionally give rise to HMBC cross peaks if the bonds between the coupling partners are arranged in w-form (see also Chapter 3.10).

3.11.3.5 Presentation of NMR Data (500 resp. 125 MHz, CDCl$_3$, δ)

Assignment	^{1}H (J)	^{13}C ($^1J_{C,H}$; $J_{C,F}$)	HMBC responses (^{13}C partners)
2	–	139.4	
3	7.84, dd (3.8, 1.2 Hz)	132.7 (170)	C-2, C-4. C-5, C=O
4	7.21, dd (4.9, 3.8 Hz)	128.9 (170)	C-2, C-3, C-5
5	7.76, dd (4.9, 1.2 Hz)	135.3 (186)	C–2, C-3, C-4, C=O
C=O	–	182.7	
=CH	6.45, s	93.5 (2.8)	C–2, C=O, =C-OH, CF$_3$
=C–OH	14.7	171.2 (36)	
CF$_3$	–	117.6 (280)	

3.12 Problem 12

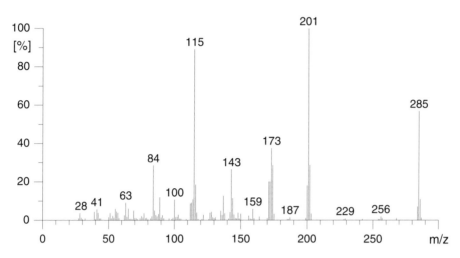

Fig. 3.12.1: Mass spectrum, EI, 70 eV

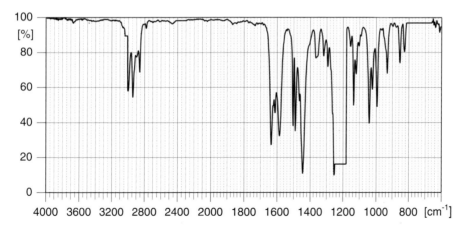

Fig. 3.12.2: IR spectrum, solvent CHCl₃, cell thickness 0.2 mm

UV spectrum (solvent ethanol):

	λ_{max}	log ε
	244	4.37
	258	4.37
	310	4.64
	343	4.84

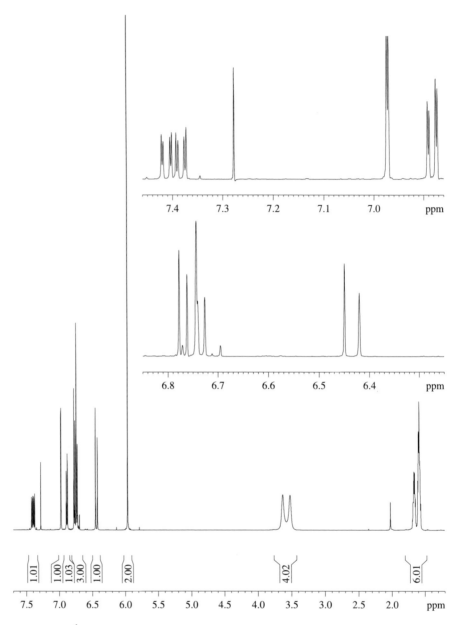

Fig. 3.12.3: ^{1}H NMR spectrum, 500 MHz, solvent CDCl$_3$

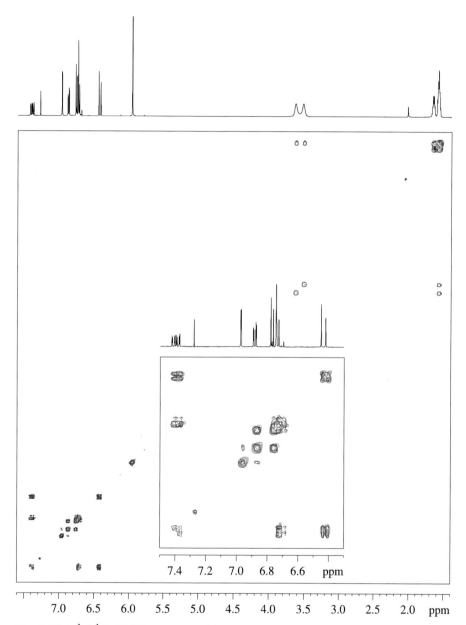

Fig. 3.12.4: ^{1}H,^{1}H COSY spectrum, 500 MHz, solvent CDCl$_3$

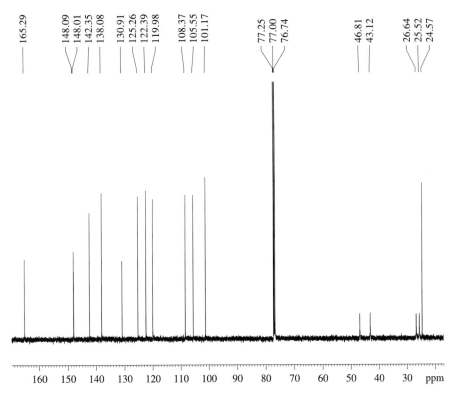

Fig. 3.12.5: ^{13}C NMR spectrum, 125 MHz, solvent CDCl$_3$

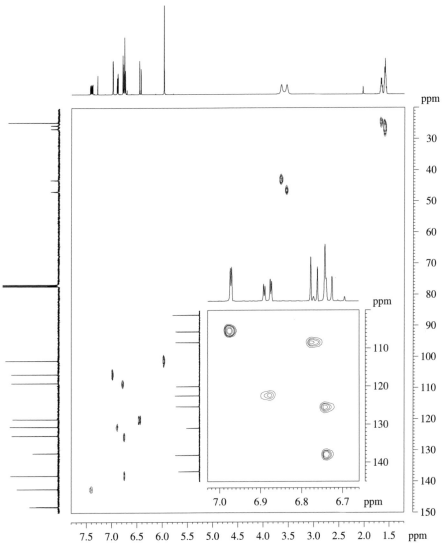

Fig. 3.12.6: $^{13}C,^{1}H$ HSQC spectrum, solvent CDCl$_3$

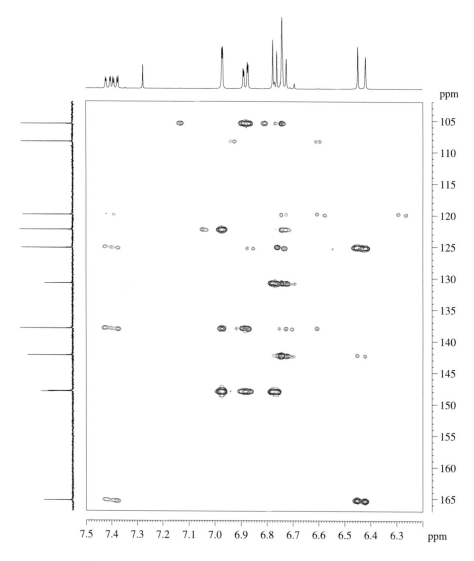

Fig. 3.12.7: $^{13}C,^{1}H$ HMBC spectrum, solvent CDCl$_3$

3.12.1 Elemental Composition and Structural Features

In the IR spectrum (Fig. 3.12.2), we note CH stretching frequencies above 3000 cm^{-1} (weak) as well as at the lower end of the standard range at 2850 cm^{-1}. The former indicate hydrogen atoms bonded to sp^2-hybridized carbon atoms, whereas the latter suggests methyl or methylene groups bonded to the hetero atom in an ether or an amine. The intense band at 1635 cm^{-1} could be due to a carbonyl group that would have to be part of a strongly delocalized π-system in order to explain its low stretching frequency.

Based on its high intensity owing to the conjugation with the C=O group, the band at 1585 cm^{-1} may be due to the C=C counterpart of this π-system. The two bands at 1500 cm^{-1} suggest a benzene ring.

The mass spectrum (Fig. 3.12.1) ends with a peak cluster centered at m/z 285. This value may be taken as the molecular mass as all differences to lower mass peaks are chemically reasonable. Since 285 is an odd number, an odd number of nitrogen atoms must be present. The general type of the spectrum is both aromatic (intensity concentrated on a few peaks in the higher mass range) and nonaromatic (according to the mass values of the fragments in the lower mass range). We note that the molecular ion predominantly loses 84 u to give the base peak at m/z 201, and we find m/z 84 as the most important peak in the low mass region.

Integration of the ^{1}H NMR spectrum (Fig 3.12.3) yields a proton ratio of 1:1:1:3:1:2:4:6 (from left), summing up to a total of 19 protons. From the integrals of the signals in the aliphatic region at δ ca. 3.6 and ca. 1.6, we assume that five methylene groups are present, the two with the higher chemical shift value being bonded to a hetero atom. This saturated heterocyclic moiety could correspond to the fragment at m/z 84, which is lost from the molecular ion in the mass spectrum leaving a predominantly aromatic remainder. With five methylene groups, we need 14 u to complete the fragment. This could correspond to the nitrogen atom already inferred from the mass spectrum. We, thus, conclude the presence of a piperidine ring. The HSQC spectrum (Fig. 3.12.6) directly gives its ^{13}C chemical shifts. The signal intensities of these methylene carbons are unusually low (Fig. 3.12.5) and the NCH$_2$ proton signals are broad. If the piperidine ring had a fast ring inversion and was free to rotate, we would find only three sharp carbon signals. On the other hand, if it was in a fixed conformation, five sharp signals would be expected in the absence of symmetry. The broadened lines indicate an intermediate rate of rotation (see also Comments). Hindered rotation of nitrogen compounds is commonly observed in amides and, sometimes, in enamines (slow rotation around the =C–N bond). Based on the IR spectrum (see above) and the signal at δ 165.3 in the ^{13}C NMR spectrum, we tentatively postulate that the piperidine ring may be attached to a carbonyl group.

Another conspicuous signal in the ^{1}H NMR spectrum is the sharp singlet at δ 5.96 of intensity 2H. As seen in the HSQC spectrum, the corresponding ^{13}C chemical shift is δ 101.2. These chemical shifts strongly suggest a methylenedioxy group.

The COSY spectrum (Fig. 3.12.4) reveals the presence of four separate proton spin systems, i.e., those of the piperidine ring, the –OCH$_2$O– group, and two systems between δ 6.4 and 7.4. The network of the shifts at δ 6.97, 6.88, and 6.77 can be easily assigned to an aromatic ring. These protons exhibit one small (1.5 Hz), one small and one large (ca. 1.5 and 8.0 Hz), and one large coupling (8.0 Hz), respectively (see Fig. 3.12.3). Such a pattern clearly indicates a 1,2,4-trisubstitution. The fourth spin system comprises the four remaining methine groups with sp^2-hybridized carbon atoms as indicated by their ^{13}C and ^{1}H chemical shifts. The doublet at δ 6.44 with a coupling constant of 14.7 Hz indicates a *trans* (*E*) disubstituted carbon-carbon double bond and can be assigned to one terminal =CH of the two conjugated double bonds. From the COSY spectrum, the chemical shift of the other proton of this double bond is δ 7.40. Its signal consists of 10 lines (including the two small ones) as a consequence of higher order effects, which must be caused by the strong couplings of the remaining two protons of the other double bond. As shown by the HSQC cross peaks 6.74/125.3 and 6.72/138.1, their ^{1}H chemical shifts are, indeed, very similar. A closer inspection of their proton signal pattern reveals a large coupling constant (J = 15.2 Hz), indicating that also this double bond is *trans* (*E*) disubstituted.

3.12.2 Structure Assembly

We may now summarize the fragments identified so far:

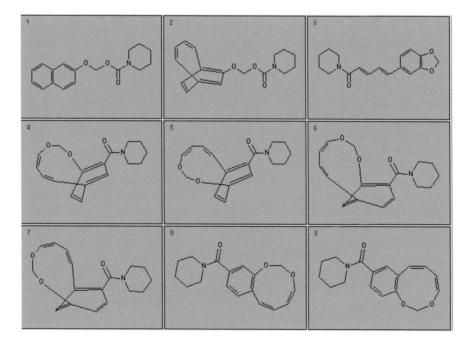

The elemental composition is calculated as $C_{17}H_{19}NO_3$ with a molecular mass of 285 u. This is in perfect agreement with the value indicated by the mass spectrum. The molecule is, thus, complete.

Fig. 3.12.8: Structures generated by Assemble 2.1

Check with Assemble 2.1

Based on the molecular formula and the fragments listed above, Assemble 2.1 generates the nine structures shown in Fig. 3.12.8 (if improbable structures are excluded). At first glance, some of them might seem surprising but their inspection is important for understanding the basic philosophy and some peculiarities of the program. The strength of a structure generator such as Assemble 2.1 is that it makes all possible combinations without having any chemical prejudice. Since it works with connectivities only, there is no possibility of constraining the structures to those having *trans* double bonds. With this constraint, only structure 3, piperine, is compatible with the input.

The two- and three-bond C,H connectivities obtained from the HMBC spectrum (Fig. 3.12.7) prove the validity of the structure found and, in addition, provide complete ^{1}H and ^{13}C signal assignments. The HMBC cross peaks of the methylene protons assign the C-1 and C-2 chemical shifts. From H-3 (δ 6.97) we find the C-1, C-5 and C-1' signals via coupling over three bonds. On the basis of the H-5/C-1' response (6.88/138.1), we obtain a further support for the assignment of C-1'. Thus, the signal assignment for the diene moiety is now straightforward. Finally, the cross peaks 7.40/165.3 and 6.44/165.3 prove, that the amide carbonyl is bound to the other end of the diene.

3.12.3 Comments

3.12.3.1 Mass Spectrum

The fragmentation of the compound follows the expected path. Initial amide cleavage yields m/z 201 and 84, the major aromatic product eliminates CO (m/z 201→173), and formaldehyde (m/z 173→143), and again CO (m/z 143→115) in succession to produce the unsaturated hydrocarbon residue $C_9H_7^+$, which further dehydrogenates. Amide cleavage is accompanied by a hydrogen transfer reaction (possibly of McLafferty type, with the double bond next to the carbonyl as hydrogen acceptor) to produce m/z 202, which also loses carbon monoxide and formaldehyde. This interpretation has been confirmed by independent experiments (not shown here).

3.12.3.2 ^{1}H and ^{13}C NMR Spectra

The interpretation of higher-order spectra is delicate, often anti-intuitive, and sometimes outright misleading. Therefore, it should be confirmed by spectrum simulation if accurate values of chemical shifts and coupling constants are required. For example, four lines would be expected in a first-order spectrum for H-3' since it has two coupling partners (the coupling over four bonds, $^4J_{1'3'}$, in dienes is < 1 Hz). The presence of ten lines (including the two small ones) is a consequence of second-order effects. The interpretation above has been confirmed by spectrum simulation.

With an exchange taking place between two or more environments of a nucleus, single lines are observed if the average lifetime of the system is short, whereas discrete lines are observed if its average lifetime is long; broad lines are observed for intermediate exchange rates. The NMR time scale is defined by the reciprocal chemical shift differences (measured in Hz) of the nuclei in different environments. Since the line frequencies depend on the magnetic field strength, this also influences the NMR time scales. In contrast to the spectrum given here, in a ^{1}H NMR spectrum recorded at 100 MHz at the same sample temperature, the two lines at δ 3.6 of the methylene protons next to the nitrogen atoms of the piperidine moiety give rise to one single broad line because the rotation around the CO–N bond at that magnetic field strength is relatively fast. Even in the high-field ^{1}H and ^{13}C NMR spectra at hand, the exchange still leads to line broadening. The prediction of the expected number of ^{13}C signals by Assemble 2.1 is based on a heuristic algorithm that cannot treat such delicate cases. In the present example, Assemble 2.1 would assume fast rotation and expect 15 signals instead of 17. To be on the safe side in such a situation, it is advisable to give a range of possible values for the number of lines (min 15, max 17).

3.12.3.3   Presentation of NMR Data (500 resp. 125 MHz, CDCl₃, δ)

Assignment	^{1}H (J)	^{13}C	Selected HMBC responses (^{13}C partners)
1		148.0	
2	–	148.1	
3	6.97, d (1.5 Hz)	105.6	C-1, C-5, C-1'
4	–	130.9	C-2, C-3, C-5
5	6.88, dd (8.0, 1.5 Hz)	122.4	C-1, C-3, C-1'
6	6.77, d (8.0 Hz)	108.4	
OCH₂O	5.96, s	101.2	C-1, C-2
1'	6.72, m	138.1	
2'	6.74, m	125.3	
3'	7.40, m	142.4	C-1', C-2', C-4', C-5'
4'	6.44, d (14.7 Hz)	120.0	C-2', C-3', C-5'
5'		165.3	
6'	3.63, m, 2H	43.1	
7'	1.58, m, 2H	26.6*	
8'	1.65, m, 2H	24.6	
9'	1.58, m, 2H	25.5*	
10'	3.52, m, 2H	46.8	

* Tentative assignments

3.13 Problem 13

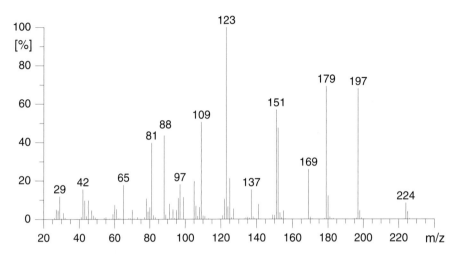

Fig. 3.13.1: Mass spectrum, EI, 70 eV

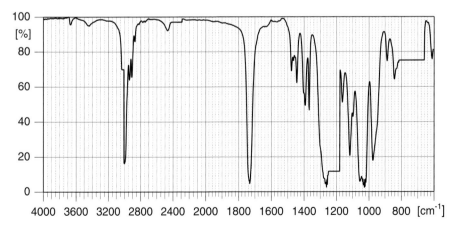

Fig. 3.13.2: IR spectrum, solvent CHCl₃, cell thickness 0.2 mm

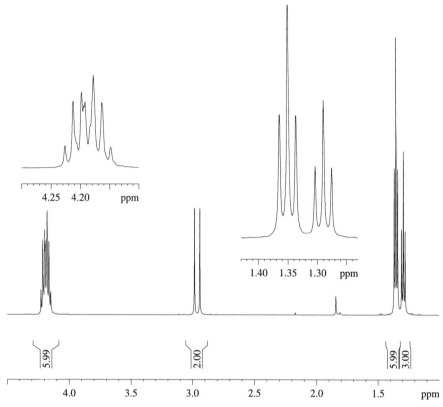

Fig. 3.13.3: ^{1}H NMR spectrum, 500 MHz, solvent CDCl$_3$

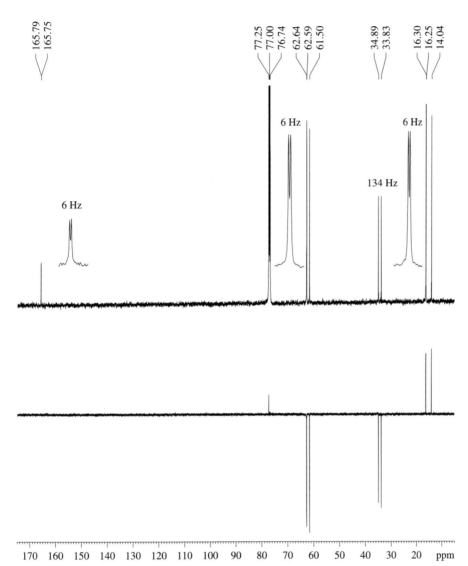

Fig. 3.13.4: ^{13}C NMR spectra, 125 MHz, solvent CDCl$_3$. Top: proton-decoupled; bottom: DEPT135

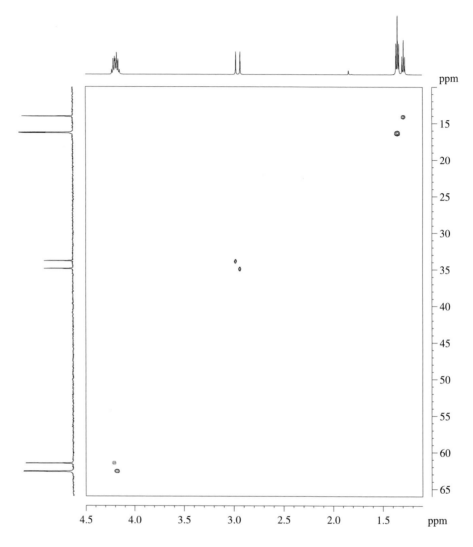

Fig. 3.13.5: $^{13}C,^{1}H$ HMBCspectrum, solvent CDCl$_3$

3.13.1 Elemental Composition and Structural Features

In the mass spectrum (Fig. 3.13.1), m/z 224 probably represents the molecular ion. The intensity distribution indicates a nonaromatic system. Loss of 45 u from M$^{+·}$ (m/z 224 → 179) and fragments at m/z 31, 45, and 88 are indicative of oxygen.

The peak at m/z 197 is accompanied by a conspicuously small ^{13}C-isotope peak of only 7% relative intensity, which allows for no more than about six carbon atoms (see also Comments to Problem 8). Heteroatoms must, therefore, largely contribute to the molecular mass.

Weak bands at ca. 3660 and 3450 cm^{-1} in the IR spectrum (Fig. 3.13.2) suggest hydroxyl stretching vibrations (free and associated, respectively), but together with the shoulder at 1630 cm^{-1}, they indicate that the sample may contain water as an impurity. A strong band at 1735 cm^{-1} shows the presence of a carbonyl group. Broad bands of high intensity around 1260 and 1040 cm^{-1} could be attributed to B–O, C–F, C–N, C–O, C=S, S=O, P=O, or P–O stretching vibrations, among which boron- or sulfur-containing functions are ruled out by the lack of natural isotope contribution to m/z 197 (at m/z 196 or 199, respectively).

The integral ratios in the ^{1}H NMR spectrum (Fig. 3.13.3) are 6:1:1:6:3 (from left), summing up to 17 protons. The high-field multiplet consists of two triplets at δ 1.35 and 1.29 (both with J = 7.0 Hz) corresponding to two and one methyl groups, respectively, which must be coupled to protons giving rise to the multiplet of 6 protons around δ 4.2. Considering their chemical shifts, these protons must be attached to oxygen atoms, thus constituting three ethoxy groups. The two remaining signals could represent two isolated methine protons. However, this contradicts the ^{13}C NMR spectrum (Fig. 3.13.4, see below).

The ten signals in the proton-decoupled ^{13}C NMR spectrum can be assigned to three CH$_3$, five CH$_2$, and two C according to the DEPT spectrum (Fig. 3.13.4, bottom). This would require a total of 19 H atoms, which contradicts the ^{1}H NMR spectrum showing only 17 H atoms. The three ethoxy groups identified there are evident from the ^{13}C NMR spectrum as the three signals each at δ 14–17 (CH$_3$) and δ 61–63 (OCH$_2$). However, the two CH$_2$ signals at δ 33–35 in the ^{13}C NMR spectrum correspond to only two protons at δ 2.9-3.0 in the ^{1}H NMR spectrum (cf. HSQC spectrum, Fig. 3.13.5). The most rational explanation of this discrepancy is the presence of heteronuclear coupling of the ^{13}C and ^{1}H atoms with a nucleus of spin I = 1/2, resulting in a doublet of the CH$_2$ group with coupling constants of 134 and 22 Hz in the ^{13}C and ^{1}H NMR spectra, respectively. Phosphorus (100% ^{31}P, I = 1/2) is the most likely candidate for such a nucleus because its presence is suggested by the strong bands in the region 1250–1050 cm^{-1} in the IR spectrum and by some features in the mass spectrum (a signal at m/z 47, the absence of a ^{13}C isotope peak for the fragment of m/z 97, and the isolated signal at m/z 65). According to the following mass balance, also the two signals at δ 166 correspond to only one C=O having a coupling to ^{31}P:

Fragment	Mass
–OCH$_2$CH$_3$	45
–OCH$_2$CH$_3$	45
–OCH$_2$CH$_3$	45
C=O	28
–CH$_2$–	14
P	31
C$_8$H$_{17}$O$_4$P	208

The missing 16 u with respect to the molecular mass of 224 must be attributed to one additional O atom because no further protons are available. The molecular formula then becomes $C_8H_{17}O_5P$.

3.13.2 Structure Assembly

Owing to its ^{13}C NMR chemical shift value and IR stretching frequency, the carbonyl group is assigned to an ester. Considering the available structural elements, it constitutes an ethyl ester moiety. Since the ^{1}H and ^{13}C chemical shifts of the methylene group (δ 2.96 and 34.3, respectively) exclude a neighboring oxygen atom, the following constitution is the only chemically meaningful possibility:

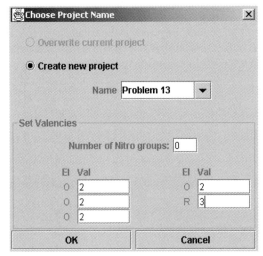

Check with Assemble 2.1

Assemble 2.1 cannot generate phosphorus compounds since it only accepts elements with a maximum valence of 4. However, it is possible to define element groups as so-called superatoms. For example, if the P=O group is found during the structure elucidation process, it can be replaced by a trivalent atom R, in the above example giving $C_8H_{17}O_4R$ as the modified molecular formula. After entering the input information, Assemble 2.1 presents the window shown in Fig. 3.13.6, where on the lower right side the valence of R is entered. With the fragments 3 x CH_3CH_2O- and one sp^3-hybridized $-CH_2-$ not bound to oxygen (the hybridization and the neighboring atom can be entered as atom tags), Assemble 2.1 finds five structures, only one of them having also a C=O group.

Fig. 3.13.6: The valence of the superatom R can be entered in the input window of Assemble 2.1

3.13.3 Comments

3.13.3.1 Mass Spectrum

Loss of 27 u from the molecular ion to give a prominent peak (shown by independent experiments to be the starting point of most of the subsequent degradations) arises by a double hydrogen rearrangement from ethoxyl, which is typical of alkyl phosphates and also rather common with many alkyl esters of carboxylic acids, especially if they have ≥ 3 C atoms in the alkyl chain. The formation of the even-numbered maximum at m/z 88 is the result of a subsequent McLafferty reaction involving transfer of one of the rearranged hydrogen atoms from oxygen onto the ester carbonyl group in a six-membered cyclic transition state and formation of an ethyl acetate ion. The fragment at m/z 65 could have been interpreted as an indication of phosphorus oxide (H_2PO_2) by itself because the absence of adjacent isotope peaks rules out the usual hydrocarbon composition C_5H_5. Similarly, alkyl phosphates yield m/z 99 (H_4PO_4) without m/z 100.

A search for characteristic ion sequences in the lower mass range, which usually is one of the first measures taken to get some general information on the compound type from the mass spectrum, does not yield any sequence with more than three members, m/z 31, 45, 59 in the oxygen series being the only perceivable one. In such cases, the appropriate conclusion is that no coherent carbon skeleton of significant length is present because consistent ion sequences depend on a carbon chain of some minimal length. Especially noteworthy is the absence of the otherwise ubiquitous fragment at m/z 39 ($C_3H_3^+$), thus practically excluding a hydrocarbon backbone of three carbon atoms or more. A missing m/z 41 ($C_3H_5^+$) is of similar significance in a nonaromatic system.

3.13.3.2 Infrared Spectrum

In our comparatively simple compound, most bands in the fingerprint region between 1500 and 1000 cm^{-1} can be rationalized. At ca. 1480 cm^{-1}, we find the deformation vibration of the methylene groups in the ethoxyl substituents on the phosphorus atom and at ca. 1445 cm^{-1} is the asymmetric deformation vibration of the methyl groups. Between these two absorption bands, we have the corresponding vibrations of the ethyl ester group. In the relatively broad band around 1400 cm^{-1}, two absorptions overlap, namely one due to deformations of the CH_2 group between the phosphorus atom and the carbonyl (at ca. 1405 cm^{-1}) and the other due to wagging vibration of the remaining methylene groups (at 1395 cm^{-1}). The symmetric deformation vibration of the methyl groups in the ethoxyl substituents on the phosphorus atom gives rise to the absorption at 1370 cm^{-1}, and the C–O–P moiety to the strong absorption around 1040 cm^{-1}. Splitting of this band into a doublet is characteristic of ethoxyl groups on phosphorus. Such a splitting is observed neither in the methoxy- nor in higher alkoxy-phosphorus compounds.

3.13.3.3 ^{1}H NMR Spectrum

The vicinal $^3J_{P,H}$ coupling constant is around 6–8 Hz and, thus, has nearly the same value as the vicinal proton-proton coupling constant. In a first-order spectrum with isochronous methylene protons, a quintet may occur for the methylene protons. In the present case, the geminal methylene protons of the two ethoxyl groups on the P atom, which are equivalent by symmetry, are diastereotopic (see also Chapter 4.3) and the predicted spin system is of the A_3MNX type. In reality, the two systems must not be considered independently since they are interlinked by the phosphorus atom (X) as a common coupling partner so that the spin system, in fact, corresponds to $A_3A_3'MM'NN'X$. Additionally, its lines overlap with the quadruplet of the ethoxycarbonyl group, which may be further split by a small long-range coupling to the phosphorus atom.

3.13.3.4 ^{13}C NMR Spectrum

Besides the methyl and methylene carbon atoms in the ethoxycarbonyl group, all ^{13}C signals are split into doublets due to ^{13}C-^{31}P couplings. The ^{13}C-^{31}P coupling constant over one bond is 134 Hz, those over two bonds are 6 and 7 Hz, and that over three bonds is 6 Hz. ^{1}H-^{1}H coupling constants generally decrease with increasing number of bonds between the coupling partners. In contrast, coupling constants between heavier nuclei often first increase with increasing distance between the nuclei, pass over a maximum, and then decrease again.

3.13.3.5   Presentation of NMR Data (500 resp. 125 MHz, CDCl$_3$, δ)

Assignment	^{1}H (J)	^{13}C ($J_{C,P}$)
CH$_3$	1.35, t (7.0 Hz)	16.3 (6 Hz)
POCH$_2$	4.18, quintet (ca. 7 Hz)	62.6 (6 Hz)
CH$_3$	1.29, t (7.0 Hz)	14.0
OCH$_2$	4.20, q (7.0 Hz)	61.5
PCH$_2$	2.96, d (21.5 Hz)	34.3 (134 Hz)
C=O	–	156.8 (6 Hz)

3.14 Problem 14

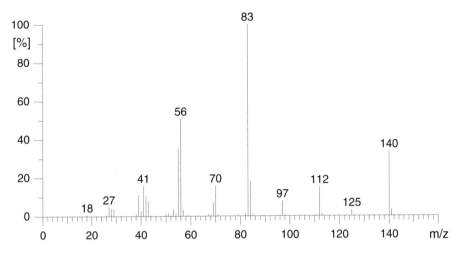

Fig. 3.14.1: Mass spectrum, EI, 70 eV

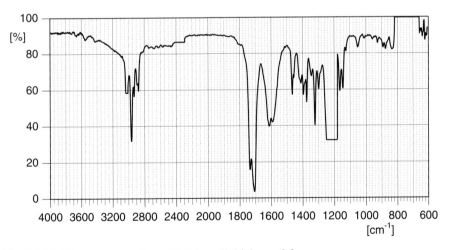

Fig. 3.14.2: IR spectrum, solvent CHCl₃, cell thickness 0.2 mm

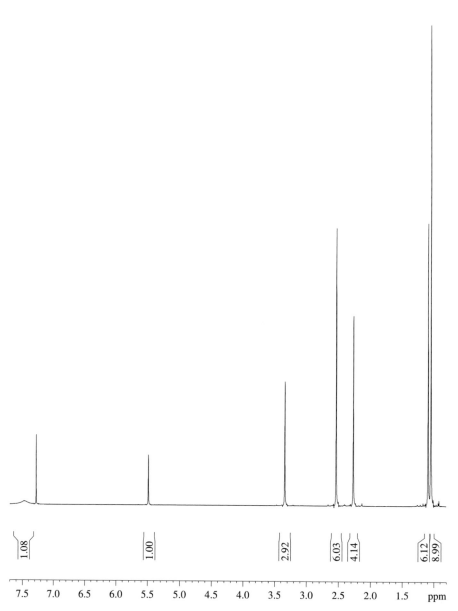

Fig. 3.14.3: ^{1}H NMR spectrum, 500 MHz, solvent CDCl$_3$

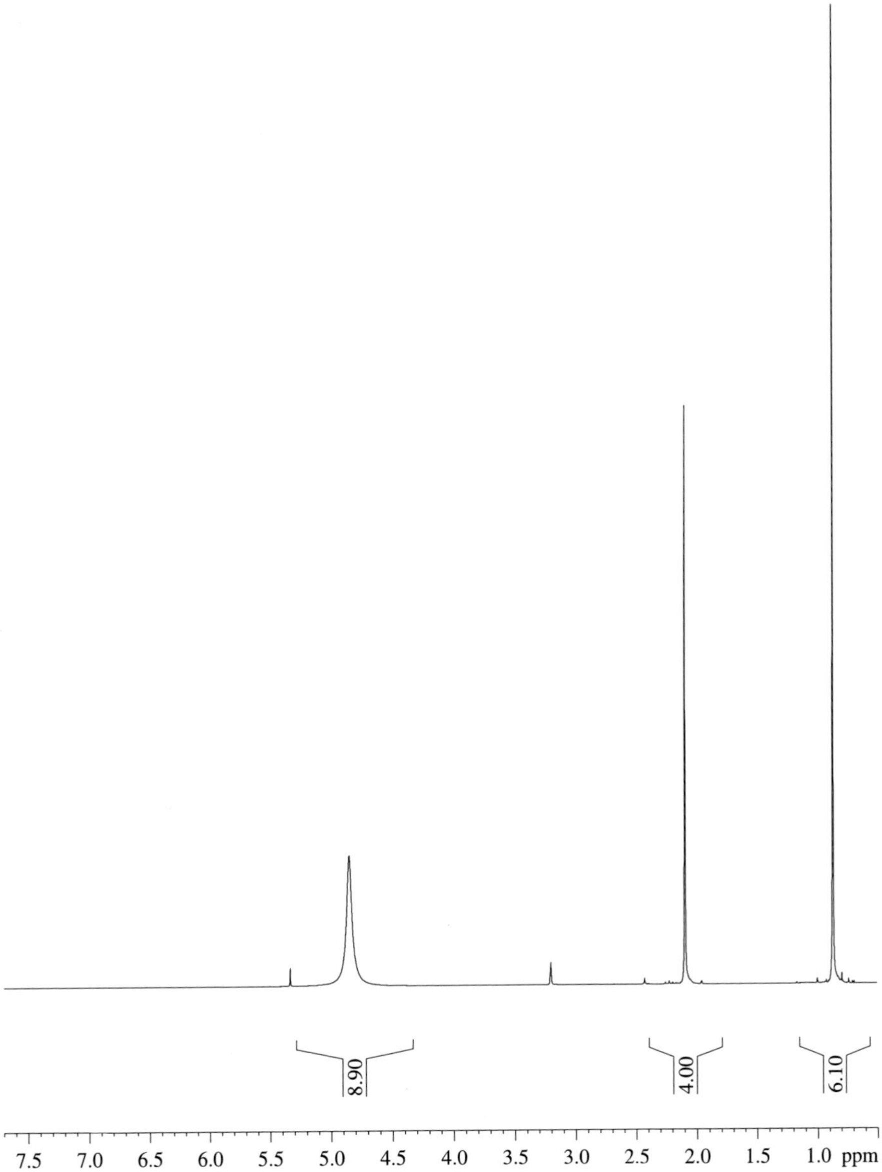

8.90

4.00

6.10

7.5   7.0   6.5   6.0   5.5   5.0   4.5   4.0   3.5   3.0   2.5   2.0   1.5   1.0 ppm

Fig. 3.14.4: ^{1}H NMR spectrum, 500 MHz, solvent CD_3OD

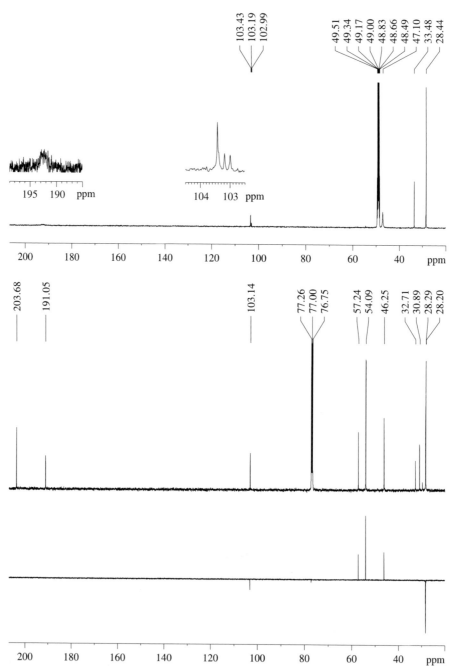

Fig. 3.14.5: ^{13}C NMR spectra, 125 MHz. Top and middle: proton-decoupled, solvent CD_3OD and $CDCl_3$, respectively; bottom: DEPT135, solvent $CDCl_3$

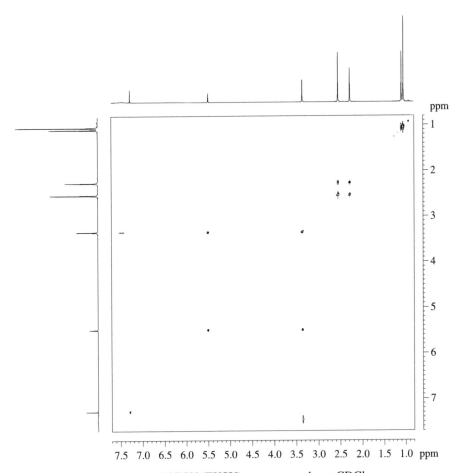

Fig. 3.14.6: Phase-sensitive NOESY (EXSY) spectrum, solvent CDCl₃

3.14.1 Elemental Composition and Structural Features

The mass spectrum (Fig. 3.14.1) ends with m/z 140, which is assumed to be the molecular mass. The first fragment peaks correspond to the loss of 15 (m/z 125), 28 (m/z 112) and 43 u (28 + 15, m/z 97), respectively. Intensity distribution and fragment masses indicate a nonaromatic unsaturated system and no evidence for nitrogen atoms.

In the IR spectrum (Fig. 3.14.2), we find a structured broad absorption band in the range of 3600–2400 cm⁻¹. Bands of this type are generally observed for ammonium compounds (N⁺–H stretching and combination bands) and for strongly hydrogen-bonded O–H compounds (O–H stretching vibrations). We need at least two oxygen atoms in order to form a strongly hydrogen-bonded OH group. Indeed, we find carbonyl stretching vibration bands at 1735 and 1705 cm⁻¹. It is not possible at this stage to interpret the strong absorption bands at about 1600 cm⁻¹.

The ^{1}H NMR spectrum recorded in CDCl$_3$ (Fig. 3.14.3) seems rather confusing. It has one broad (δ 7.47) and six sharp singlets (δ 5.49, 3.34, 2.53, 2.26, 1.09, and 1.04) with intensity ratios of 1:1:3:6:4:6:9 (from left) corresponding to 30 H atoms in total, which contradicts the assumed molecular mass. With CD$_3$OD as solvent (Fig. 3.14.4), we only find three signals, one broad (δ 4.84) and two sharp singlets (δ 2.10 and 0.87), with an intensity ratio of 9:4:6. The low-field signal thereby corresponds to the OH proton of the solvent and contains all exchangeable protons of the sample as well as of nondeuterated methanol and water from the solvent, if present. We notice that the chemical shifts of the two sharp singlets at δ 2.33 and 1.16 are close to the corresponding shifts of two pairs of signals in CDCl$_3$ (δ 2.53/2.26 and 1.09/1.04). The signals at δ 7.47, 5.49, and 3.34, on the other hand, do not appear in CD$_3$OD. In this solvent, exchangeable protons are almost quantitatively substituted by deuterium since its molecular concentration is much higher than that of the sample. Therefore, we may assume that the protons corresponding to the signals that do not appear in CD$_3$OD are exchangeable and, thus, included in the signal at δ 4.84. The different number of signals suggests that the compound under study exists as different species in the two solvents. The unusual intensity ratio of the signals in CDCl$_3$ would require a much too large number of protons (30H), which cannot be accommodated with the given molecular mass (140 u). Therefore, we assume the presence of two slowly exchanging tautomers or conformers in CDCl$_3$, whereas in CD$_3$OD only one of them occurs.

Comparison of the ^{13}C chemical shifts measured in CDCl$_3$ and CD$_3$OD (Fig. 3.14.5) allows us to assign the lines to the two different forms. We note that the minor component in CDCl$_3$ corresponds to the species that appears in methanol. Thus, the major component (ca. 2/3) in CDCl$_3$ (see DEPT135 spectrum, Fig. 3.14.5, bottom) shows signals for one C=O (δ 203.7), two CH$_2$ (δ 57.2 and 54.1), and one CH$_3$ (δ 28.2), whereas the minor component (ca. 1/3) exhibits one signal each for C=O (δ 191.1), sp^2-CH (δ 103.1), CH$_2$ (δ 46.3), and CH$_3$ (δ 28.3).

3.14.2 Structure Assembly

The following structural fragments have been found for the form dominating in CDCl$_3$:

Structural fragment	Mass
2 CH$_3$ (equivalent by symmetry)	30
2 CH$_2$ (equivalent by symmetry)	28
1 CH$_2$	14
1 C	12
1 C=O	28
1 O	16

These elements sum up to C$_7$H$_{12}$O$_2$ (128 u). The difference to the molecular mass of 140 u corresponds to one carbon atom. This must be equivalent by symmetry with one of the nonprotonated carbon atoms found above since no further signals are available in the ^{13}C NMR spectrum.

Check with Assemble 2.1

With the molecular formula of $C_8H_{12}O_2$, Assemble 2.1 generates 208 557 structures if improbable fragments are forbidden. Only eight structures are generated (see Fig. 3.14.7) if the following information is entered: two sp^3-CH$_3$, two sp^3-CH$_2$, at least 1 C=O, and only 5 signals in the ^{13}C NMR spectrum. A keto-enol tautomerism is only possible with structures 7 and 8. The latter can be excluded since the CH$_3$ group appears as a singlet in the ^{1}H NMR spectrum. Hence, compound 7 (dimedone) must be the correct solution.

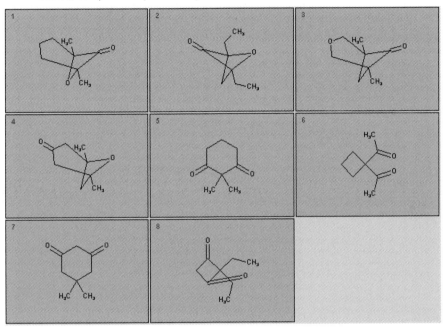

Fig. 3.14.7: Structures generated with Assemble 2.1

3.14.3 Comments

3.14.3.1 Mass Spectrum

Loss of 28 mass units from the molecular ion must be ascribed to decarbonylation, probably involving a ring contraction reaction. If it were due to loss of ethylene, one would expect it to be accompanied by loss of an ethyl radical ($\Delta m = 29$). Formation of the base peak at m/z 83 (and subsequent decarbonylation to m/z 55) is visualised according to the typical cyclohexanone reaction sequence **a → b → c → d**.

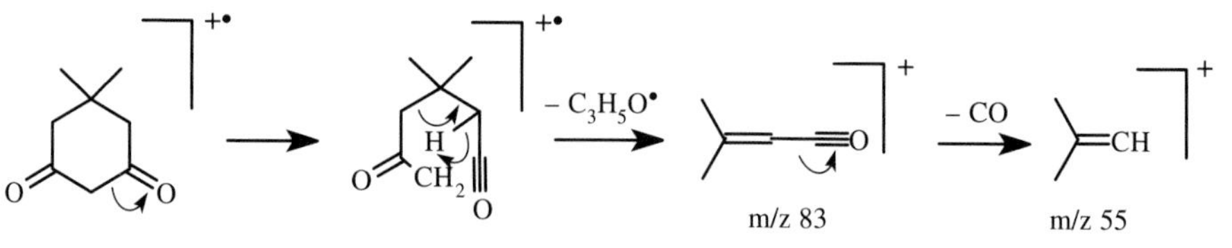

m/z 83   m/z 55

a   b   c   d

Formation of the prominent peak at m/z 56 could be the result of a *retro*-Diels–Alder reaction of the enols **II A** and **II B** (see Section 3.14.3.3) according to the following scheme:

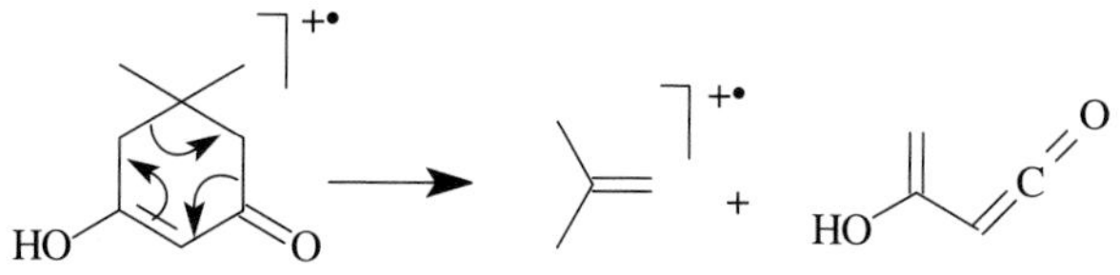

or else of the elimination of two neutral ketene molecules, which are particularly good leaving groups in gas phase reactions:

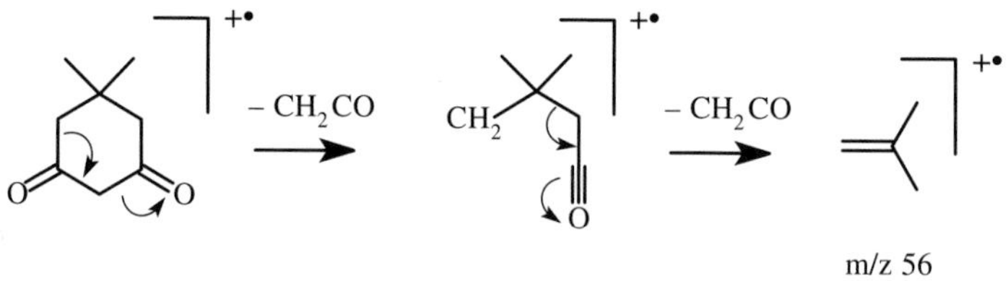

m/z 56

3.14.3.2 Infrared Spectrum

The IR spectrum corroborates the presence of a mixture of keto and enol forms in chloroform. In the keto form, the stretching vibrations of the two carbonyl groups are coupled, leading to a double band at 1735 and 1705 cm^{-1} (asymmetric and symmetric vibrations). In the enol, it is not possible to separate the various CC and CO vibrations; the strong bands around 1600 cm^{-1} are rather to be ascribed to the conjugated chromophore as a whole. In 2-monosubstituted or nonsubstituted 1,3-cyclohexadiones, often two bands are observed for the enol. One is around 1640 cm^{-1} and corresponds to the free form. The other band at about 1600 cm^{-1} is assigned to a dimer of the following type:

In chloroform, the enolized dimedone exists predominantly as a dimer, the absorption of the monomer at 1640 cm^{-1} not being discernible. Furthermore, if appreciable amounts of the free enol were present, we would expect a moderately sharp band for the O–H stretching vibration above 3000 cm^{-1}.

3.14.3.3 ^{1}H and ^{13}C NMR Spectra

There are three possible tautomeric forms of dimedone having reasonable stability:

The postulated equilibrium of these three tautomeric forms is directly proven by the phase-sensitive NOESY (nuclear Overhauser enhancement and exchange spectroscopy or EXSY, exchange spectroscopy) measurement (Fig. 3.14.6). This experiment yields positive and negative cross peaks. For relatively small molecules ($M_r < 1000$ u), the NOE (nuclear Overhauser effect) responses indicating steric proximities (< 5 Å) are positive, whereas the signals characteristic of exchange are negative and have the same sign as the diagonal signals of the two-dimensional map. During the mixing time (τ_{mix} = 800 ms), not only NOE responses build up but also magnetization transfer takes place between the exchanging protons. In the EXSY spectrum, there are cross peaks between the pairs of the methylene (δ 2.53/2.26) and methyl signals (δ 1.04/1.09) corresponding to different tautomers. A threefold exchange among the protons of the OH, =CH, and one of the CH$_2$ groups can be obtained from the δ 7.47/5.49/3.34 EXSY cross peaks. The intensities of the cross peaks depend on the mixing time. The exchange rate constant can be determined from their volume integrals obtained for different mixing times.

In CDCl$_3$ as solvent, the exchange between **I** and **II A** or **II B** is slow on the NMR time scales but fast between **II A** and **II B** so that the corresponding nuclei become magnetically equivalent (see Chapter 4.3). In solvents without exchangeable deuterium, we would, thus, observe five ^{13}C and three ^1H NMR signals for **I** and five ^{13}C and four ^1H NMR signals for **II A / II B**. Only nine of the ten predicted signals are discernible in the ^{13}C NMR spectrum (solvent $CDCl_3$) since the methyl groups of **I** and **II A / II B** are almost equivalent (δ 28.3 and 28.2). In $CDCl_3$, the chemical shift of δ 191.1 corresponds to the mean value of the shifts for =C–O and C=O in **II A / II B**. The signal at δ 103.1 can be assigned to the alkene =CH carbon atom of these tautomers.

In CD_3OD, all protons involved in the tautomeric reactions, i.e., the alkene and the hydroxyl protons of **II A / II B**, are replaced by deuterium. Since the population of the keto form **I** is very low in this solvent, no corresponding signals are detectable in the NMR spectra.

The ^{13}C NMR spectrum recorded in CD_3OD is complicated by two different facts. First, the alkene methine is now predominantly deuterated. Deuterium having a spin quantum number of $I = 1$ leads to a splitting of the signal of the coupling partners into three lines with relative intensities of 1:1:1. This coupling is, of course, not influenced by proton broad-band decoupling. The coupling constant is smaller than the corresponding C–H coupling constant by a factor of 6.514 as demanded by the gyromagnetic constants of proton *versus* deuterium. The spectrum shows a significantly higher intensity of the low-field line of the triplet (at δ 103.43). This can be explained by the presence of a small amount of nondeuterated sample. Owing to the isotope effect of deuterium, its CH signal is at lower field (δ 103.43) than the CD signal (δ 103.19). Since the relaxation times of the nondeuterated and deuterated CH are expected to be significantly different, it is not possible to directly interpret the signal intensities. Another surprising feature of the ^{13}C NMR spectrum in CD_3OD is the low intensity of the signal at ca. 192.5 ppm as a consequence of its large line width. This can be explained by the limited exchange rate between **II A** and **II B** so that the =C–O and C=O signals are not fully averaged. For the same reason, the CH_2 signal at δ 47.1 is also broadened; however, the line width is considerably smaller. This indicates that the chemical shift difference between the two methylene carbon signals for **II A / II B** is smaller than that between the =C–O and C=O signals.

3.14.3.4   Presentation of NMR Data (500 resp. 125 MHz, CDCl$_3$, δ)[a]

Assignment	Keto Form		Enol Form	
	^1H	^{13}C	^1H	^{13}C
$(CH_3)_2$	1.04	28.2	1.09 (0.87)	28.3 (28.4)
C	–	30.9	–	32.7 (33.5)
$(CH_2)_2$	2.53	54.1	2.26 (2.10)	46.3 (47.1)
CH_2	3.34	57.2	–	–
=CH	–	–	5.49[b]	103.1 (103.2)
C=O	–	203.7	–	191.1 (ca. 192.5)
OH	–	–	7.47[b]	–

[a] Values in parentheses measured in CD_3OD.
[b] Because of exchange with the solvent, no signals detected in CD_3OD. The corresponding protons appear at δ 4.84 as CD_3OH.

3.15 Problem 15

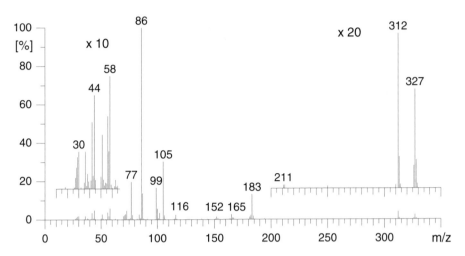

Fig. 3.15.1: Mass spectrum, EI, 70 eV

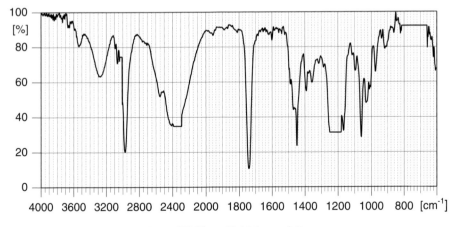

Fig. 3.15.2: IR spectrum, solvent CHCl₃, cell thickness 0.2 mm

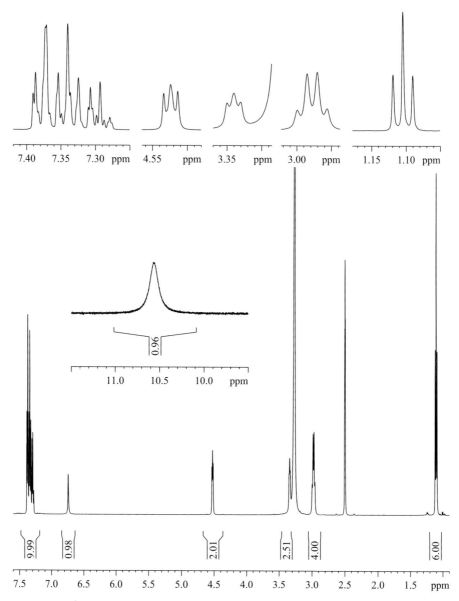

Fig. 3.15.3: ^{1}H NMR spectrum, 500 MHz, solvent DMSO-d$_6$, 320 K

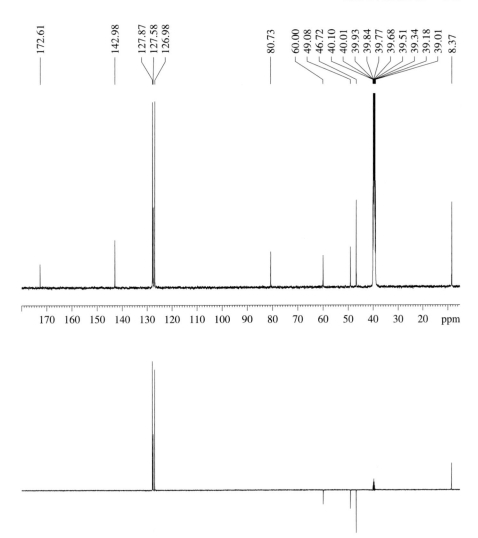

Fig. 3.15.4: ^{13}C NMR spectra, 125 MHz, solvent DMSO-d$_6$, 320 K. Top: proton-decoupled; bottom: DEPT135

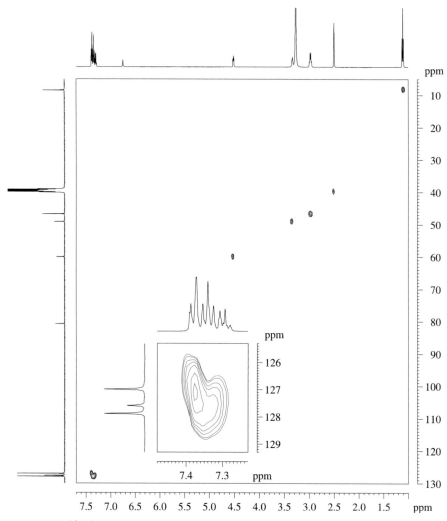

Fig. 3.15.5: ^{13}C,^{1}H HSQC spectrum, solvent DMSO-D$_6$, 320 K

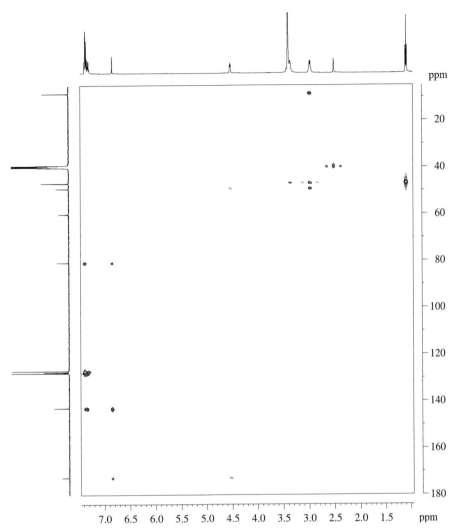

Fig. 3.15.6: $^{13}C,^{1}H$ HMBC spectrum, solvent DMSO-d_6, 300 K

3.15.1 Elemental Composition and Structural Features

In the mass spectrum (Fig. 3.15.1), m/z 327 appears to represent the molecular ion since all peaks at lower masses show chemically reasonable mass differences relative to this signal. The odd mass value indicates the presence of an odd number of nitrogen atoms, which is corroborated by a small but significant fragment at m/z 30. Weak signals at m/z 36 and 38 of proper intensity ratio indicate HCl. Since no chlorine atom is evident from what is assumed to be the molecular ion (no signal of about 30% relative intensity at m/z 329), the possibility of dealing with the hydrochloride of an amine must be considered, unless one assumes HCl to be present as an impurity. The fragment series m/z 51, 77, 105, 183 (= 105 + 78) and m/z 30, 44, 58, 72, 86, respectively, reveal a mixed aromatic and nonaromatic nature of the unknown compound.

The two bands at 3540 and 3280 cm^{-1} in the IR spectrum (Fig. 3.15.2) indicate a hydroxyl group (OH stretching vibrations of free and hydrogen-bonded OH group). An NH stretching vibration at 3540 cm^{-1} is very unlikely and would give rise to a sharp absorption. The broad band at 2400 cm^{-1} is taken as evidence corroborating the presence of an ammonium salt (see Comment). The absorption at 1740 cm^{-1} indicates a carbonyl group.

The relative signal intensities in the ^{1}H NMR spectrum measured in DMSO-d$_6$ (Fig. 3.15.3) are 1:10:1:2:2:4:6 (from left) and correspond to 26 H atoms in total. The signal of the solvent appears at δ 2.50, the strong singlet at ca. δ 3.3 being due to water, which usually occurs because DMSO is very hygroscopic. The spectrum was taken at elevated temperature (320 K) since at room temperature the signal of water overlaps with that at δ 3.34. Owing to the reduction in the relative amount of hydrogen-bonded species, the chemical shifts of OH and NH protons generally decrease with increasing temperature, whereas the chemical shifts of the CH protons remain practically unaffected. A comparison of the spectra obtained at 300 K (not shown) and 320 K also shows upfield shifts of the signals at δ 10.57 and 6.74 (each 1H). The quartet at δ 2.98 (4H) and the triplet at δ 1.11 (6H) indicate the presence of two equivalent ethyl groups, which must be attached to the nitrogen as shown by the chemical shift of their CH$_2$ groups. The triplet-like symmetric signal pattern of the multiplets at δ 4.52 (2H) and 3.34 (2H) must belong to two neighboring methylene groups (*AA'XX'* spin system). Because of their chemical shifts, they must be attached to electronegative atoms. The signals in the range of δ 7.45–7.25 with an intensity of 10H indicate the presence of two monosubstituted phenyl groups. Although they correspond to a higher-order spectrum, three multiplets with characters of a doublet, triplet, and triplet (of lower intensity) are discernible.

Only 10 nonisochronous carbon atoms are discernible in the proton-decoupled ^{13}C NMR spectrum (Fig. 3.15.4, top). According to the DEPT135 (Fig. 3.15.4, bottom) and HSQC spectra (Fig. 3.15.5), they can be assigned to one CH$_3$, three CH$_2$, three aromatic =CH, and three quaternary C atoms. The integration in the ^{1}H NMR spectrum indicates that some of them must be present twice. For example, only four signals appear in the aromatic region (for 3 CH and 1 C), which corroborates the presence of two symmetrically equivalent monosubstituted benzenes. In addition to the carbon atoms to be expected from the functional groups found above, we recognize a quaternary *sp*^{3}-hybridized carbon atom, which according to its chemical shift of δ 80.7 must be attached to a hetero atom. Also the C=O carbon is attached to a hetero atom as shown by its chemical shift of δ 172.6.

The structural elements are shown below, their mass adding up to 347 u. If, in fact, the sample is an ammonium chloride and m/z 327 represents the molecular ion of the corresponding base, then the complete molecular mass is 327 + 36 (HCl) = 363 and the difference to the mass balance above is 16 u, which must be attributed to one oxygen atom. The molecular formula would then be C$_{20}$H$_{25}$NO$_3$·HCl, showing nine double bond equivalents, which are taken care of by two phenyl groups and one carbonyl.

Structural elements	Mass	Structural elements	Mass
2 (phenyl ring)	154	–N–H	15
C (carbon)	12	–O–H	17
–CH$_2$–CH$_2$–	28	–C=O	28
2 –CH$_2$CH$_3$	58	–Cl	35

3.15.2 Structure Assembly

The two ethyl groups must be attached to the nitrogen atom because the ^{13}C and ^{1}H chemical shifts of their methylene groups require a hetero atom next to them and because they can only become equivalent if they are bonded to the same hetero atom. An amide carbonyl group is excluded owing to its IR stretching frequency and the required basicity of the compound. Since the IR spectrum also contradicts an acid, an ester is the only possibility left.

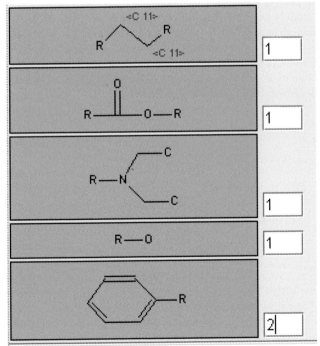

Fig. 3.15.7: Fragments entered to Assemble 2.1

Check with Assemble 2.1

The molecular formula $C_{20}H_{25}NO_3$ and the fragments shown in Fig. 3.15.7 yield the two possible isomers of Fig. 3.15.8. No carboxylic acid is generated since the input of the two fragments –COO– and –OH demands three different oxygen atoms. However, if –OH is entered as an atom constraint and not as a fragment, Assemble 2.1 additionally generates an acid.

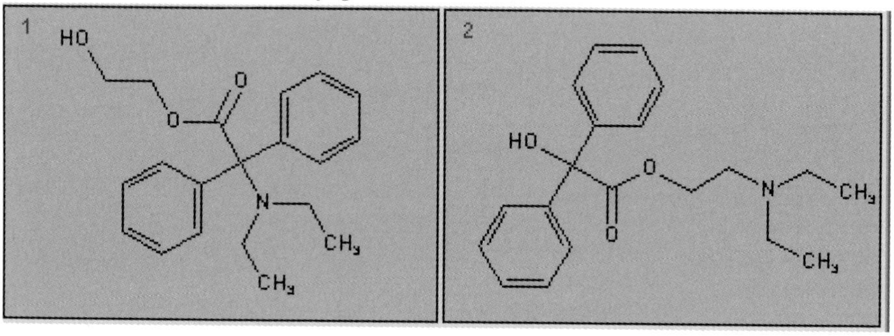

Fig. 3.15.8: Output of Assemble 2.1

The same solutions are obtained by the following reasoning: The N atom must be attached to the –CH$_2$CH$_2$– group to explain the chemical shift of a third methylene group and the appearance of the dominant even-mass base peak at m/z 86 in the mass spectrum. The phenyl groups are terminal functions and can only be bound to the quaternary C atom because equivalence of the groups and connection to sp^3-hybridized carbon are required. In addition, the quaternary C atom must carry the hydroxyl group in order to give the mass of the aromatic moiety of 183 u and explain the aromatic fragment series of m/z 183, 105, 77, 51. The elimination of benzene (m/z 183 → 105) requires a mobile hydrogen atom and the subsequent transition of m/z 105 → 77 must be due to a decarbonylation to yield a phenyl cation. To complete the molecule, the diphenylcarbinyl and the diethylamine part must be connected via CO$_2$ resulting in two possible constitutions, **I** and **II**:

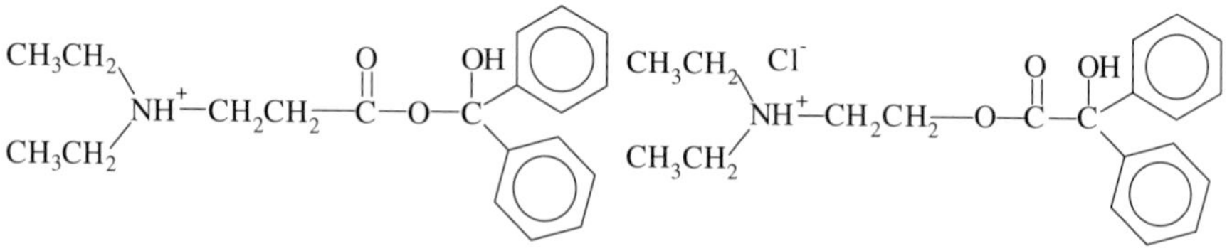

I   **II**

A decision in favour of **II** can easily be made by inspecting the two- and three-bond responses in the HMBC spectrum measured at 300 K (Fig. 3.15.6). The cross peaks of the OH signal at δ 6.74 show correlations over two bonds to the sp^3-carbon at δ 80.7, and over three bonds to the C$_{ipso}$ carbons at δ 143.0 and to the C=O at δ 172.6, which is only possible in case of structure **II**. The fragments at m/z 116 and 99 in the mass spectrum provide further arguments in favor of **II** (see Comments).

3.15.3 Comments

3.15.3.1 Mass Spectrum

Ammonium salts are not significantly volatile. Those of amines with strong acids upon heating usually yield the spectrum of the free base along with that of the free acid, the latter frequently appearing with low intensity. In this case, the weak signals of HCl are easily recognized because the mass range m/z 35–37 is usually empty. Alkyl ammonium halides can give products of Hoffmann degradations and analogous reactions, while complex anions are generally difficult or impossible to identify. As nitrogen is an excellent carrier of the positive charge, the spectrum of aliphatic amines is normally dominated by the largest nitrogen-containing fragment produced by an α-cleavage, here m/z 86:

The signal at m/z 99 can be explained by a McLafferty rearrangement involving the ester carbonyl group, where the nitrogen containing part carries the charge. Neither this rearrangement nor the fragment at m/z 116 would be possible with **I**.

3.15.3.2 Infrared Spectrum

Broad bands in the range of 3000–2000 cm^{-1} are indicative of N-H stretching and combination vibrations of ammonium salts, with the main maximum appearing below 2500 cm^{-1} only if tertiary amines are involved.

3.15.3.3 ^{1}H NMR Spectrum

The chemical shift of ammonium NH protons generally lies in the range of δ 6–8. Because of the formation of strong hydrogen bonds to the sulfoxide group, NH (and OH) protons have especially large chemical shifts in DMSO (here, δ 10.57). The signals are often broad because of the ^{14}N-^{1}H coupling, which is partially decoupled through the quadrupole relaxation of the nitrogen nuclear spin. The exchange rate between ammonium NH protons and other exchangeable protons is usually slow on the NMR time scale. In the present case, however, a slow exchange of NH with OH protons (of the sample and of H$_2$O from the solvent) should also be considered, which would contribute to the line broadening and, in addition, eliminate the splittings due to the vicinal NH–CH$_2$ couplings.

In the case of a slow exchange of the ammonium hydrogen atoms, the nitrogen inversion is also slow and the geminal CH$_2$ protons of the ethyl group are not equivalent so that an *ABX$_3$* spin system is expected. Due to symmetry reasons, the spin systems of the two ethyl groups are identical. In the actual spectrum, the signal of the NCH$_2$ protons is broadened, which can be the result of their nonequivalence and/or of the vicinal CH$_2$–NH$^+$ coupling. In both cases, the line broadening indicates a finite exchange rate, which partially obscures these effects. The exchange rate can be reduced by recording the spectrum with a strong acid, such as trifluoroacetic acid, as solvent. However, the compound at hand spontaneously reacts with trifluoroacetic acid by forming an ester (spectrum not shown).

3.15.3.4   Presentation of NMR Data (500 resp. 125 MHz, DMSO-d$_6$, 320 K, δ)

Assignment	^{1}H	^{13}C	Selected HMBC responses (^{13}C partners)
(CH$_3$)$_2$	1.11, t, 6H	8.4	N(CH$_2$)$_2$
N(CH$_2$)$_2$	2.98, q, broad, 4H	46.7	CH$_3$, NCH$_2$
NH	10.57, broad, 1H	–	–
NCH$_2$	3.34, m, 2H	49.1	N(CH$_2$)$_2$
OCH$_2$	4.52, m, 2H	60.0	NCH$_2$, C=O
C=O	–	172.6	
C	–	80.7	
OH	6.74, s, 1H	–	C, C$_{ipso}$, C=O
C$_{ipso}$	–	143.0	
CH$_{ortho}$	7.38, m, 4H	127.0	
CH$_{meta}$	7.34, m 4H	127.9	
CH$_{para}$	7.29, m, 2H	127.6	

3.16 Problem 16

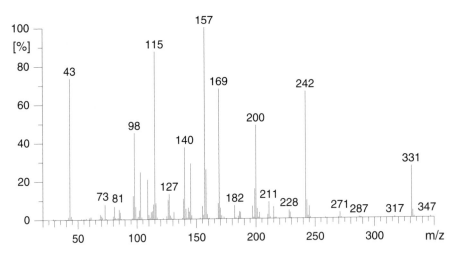

Fig. 3.16.1: Mass spectrum, EI, 70 eV

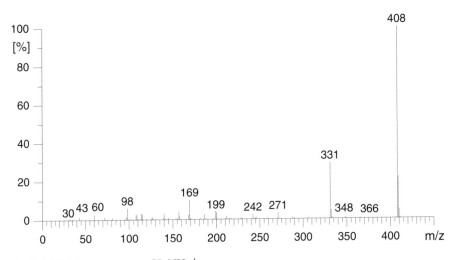

Fig. 3.16.2: Mass spectrum, CI, NH$_4^+$

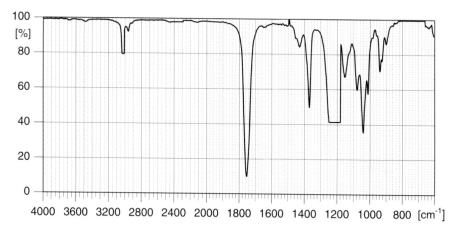

Fig. 3.16.3: IR spectrum, solvent CHCl₃, cell thickness 0.2 mm

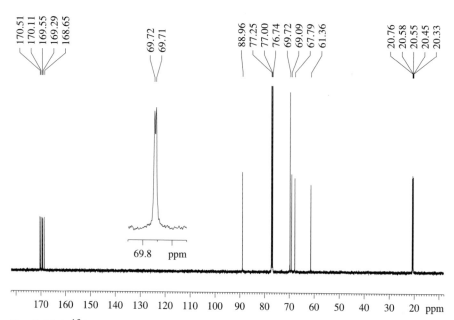

Fig. 3.16.4: ^{13}C NMR spectrum, 125 MHz, solvent CDCl₃

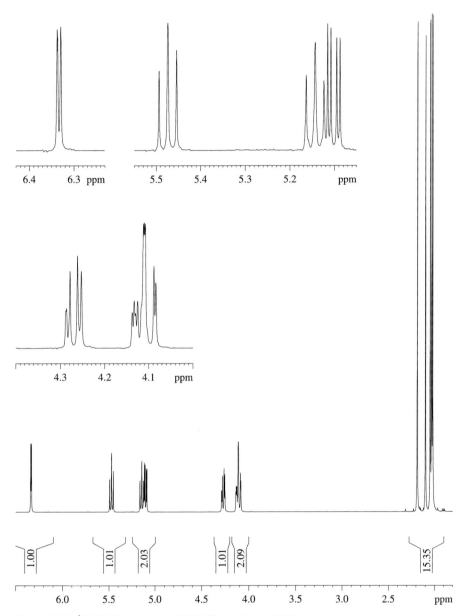

Fig. 3.16.5: ^{1}H NMR spectrum, 500 MHz, solvent CDCl$_3$

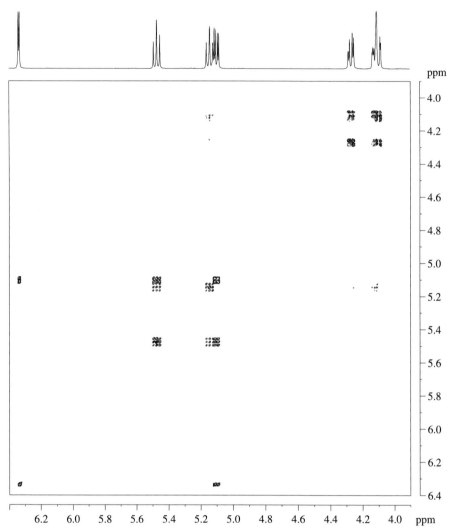

Fig. 3.16.6: ^{1}H,^{1}H COSY spectrum, 500 MHz, solvent CDCl$_3$

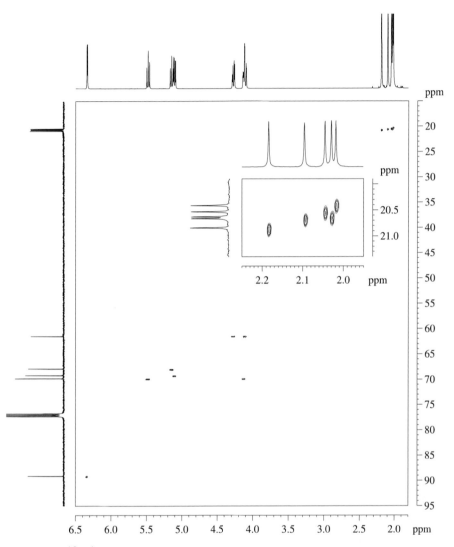

Fig. 3.16.7: ^{13}C,^{1}H HSQC spectrum, solvent CDCl$_3$

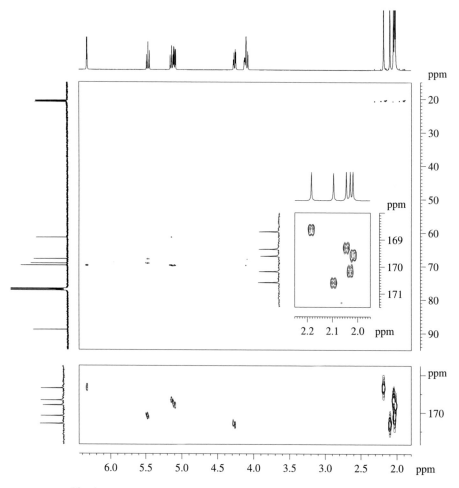

Fig. 3.16.8: ^{13}C,^{1}H HMBC spectrum, solvent CDCl$_3$

3.16.1 Elemental Composition and Structural Features

In the IR spectrum (Fig. 3.16.3), the carbonyl stretching vibration band at 1755 cm^{-1} together with two prominent bands in the region of 1300–1000 cm^{-1} (possibly C–O–C stretching vibrations, recognizable despite the solvent absorption) suggest that the compound is an ester. There are no other characteristic absorptions that could provide significant information at this stage.

The last signal in the EI mass spectrum (Fig. 3.16.1) is at m/z 347. This peak is probably not the molecular ion since the mass difference to the next fragment (m/z 331) is 16 u (see Comments). If the molecular ion is not evident, a chemical ionization (CI) spectrum will help. In the present case, the CI spectrum (Fig. 3.16.2) with ammonia as reactant gas shows m/z 408 for [M + NH$_4$]$^+$, giving a molecular mass of 390. The fragments at m/z 347 and 331, thus, correspond to $[M - 43]^+$ and $[M - 59]^+$. This is in line with the rather intense peak at m/z 43 without any ignificant amount of m/z 41, which suggests the presence of an acetyl group in the molecule.

Integration of the ^{1}H NMR spectrum (Fig. 3.16.5) yields intensity ratios of 1:1:2:1:2:3:3:9 (from left) summing up to a total of 22 H atoms in the molecule. The five signals in the range of δ 2.2–2.0 corresponding to 15 protons fit in well with the acetyl methyl groups already inferred from the mass spectrum. As all five methyl groups have very similar chemical shift values, we suspect the presence of five acetyl groups.

In the ^{13}C NMR spectrum (Fig. 3.16.4), we indeed find five carbonyl signals of very similar chemical shifts (δ 170.5–168.6). These values exclude ketones and are in very good agreement with the presence of esters as was suspected from the IR, mass, and ^{1}H NMR spectra. The five acetyl methyl carbon atoms lead to signals in the very narrow range of δ 20.8–20.3. The corresponding pairs of the ^{13}C and ^{1}H chemical shifts can easily be read from the HSQC spectrum (Fig. 3.16.7). There are five more signals in the region of δ 89.0–61.3, i.e., with chemical shift values characteristic of carbon atoms bound to oxygen. The HSQC spectrum indicates one CH_2 (δ 61.4) and five CH groups and proves that all of the protons are bound to carbon atoms. The ^{13}C and ^{1}H chemical shifts of δ 89.0 and 6.33, respectively, suggest that the corresponding CH group must be attached to two oxygen atoms.

3.16.2 Structure Assembly

The following structural fragments have been found so far:

$$5 \ CH_3-COO-$$
$$1 \ CH_2-(O)-$$
$$4 \ CH-(O)-$$
$$1 \ (O)-CH-(O)$$

We need at least one ether oxygen atom to accommodate the above fragments. The molecular formula thus becomes $C_{16}H_{22}O_{11}$ (390 u).

Check with Assemble 2.1

With the molecular formula of $C_{16}H_{22}O_{11}$, the requirement of 16 signals observed in the ^{13}C NMR spectrum, and the fragments shown in Fig. 3.16.9, Assemble generates the seven structures shown in Fig. 3.16.10. The short notation of the atom tags, <sp3> and <Osp311> or <Osp322> (entered in the corresponding dialog box of the drawing program), in Fig. 3.16.9 indicates that the corresponding C atoms are sp^3-hybridized (shown by their NMR chemical shifts) and that they are substituted by one or two sp^3-hybridized oxygen atoms, respectively.

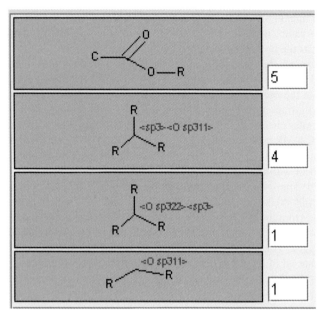

Fig. 3.16.9: Fragments entered to Assemble 2.1

All structures in Fig. 3.16.10 have one ring as the molecular formula corresponds to six double bond equivalents, while only five carbonyl groups and no C=C double bonds are present.

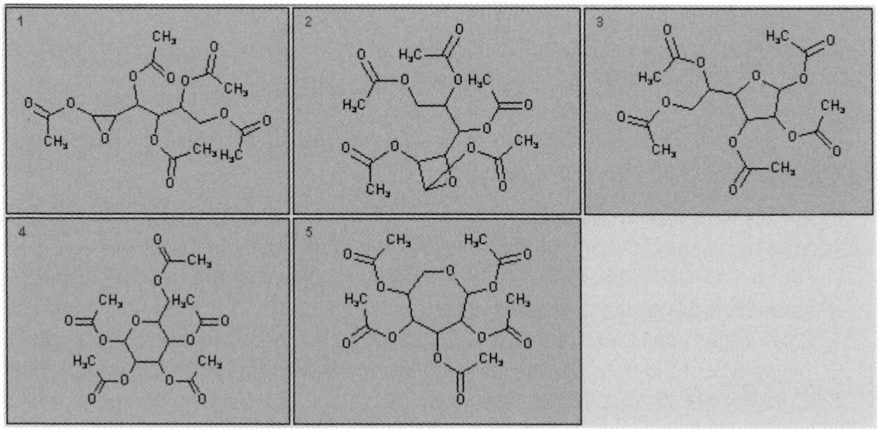

Fig. 3.16.10: Output of Assemble 2.1

The proton-proton connectivities can be derived from the COSY spectrum (Fig. 3.16.6, see also the ^{1}H NMR spectrum in Fig. 3.16.5). The doublet of the (O)-CH-(O) methine at δ 6.33 (J = 3.8 Hz) is a good starting point because it has only one coupling partner. Its cross peak leads to the signal at δ 5.10, from which we subsequently reach the signals at δ 5.47, δ 5.14, and δ 4.12. This last signal has two additional coupling partners at δ 4.27 and 4.10. According to the HSQC spectrum, these protons belong to a methylene group. The pieces of information found so far can be summarized as follows:

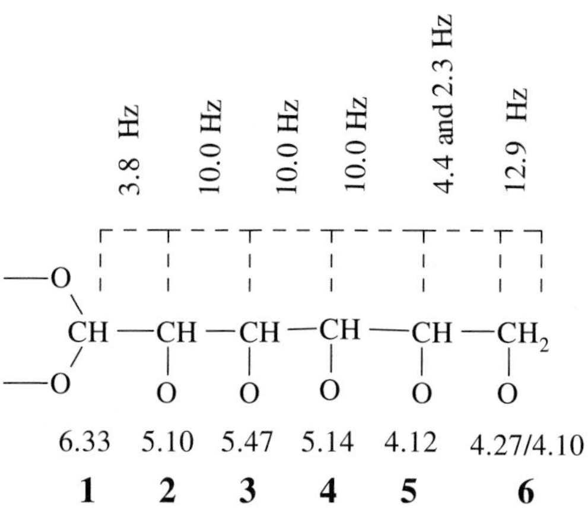

All solutions shown in Fig. 3.16.10 are compatible with this sequence but not with the chemical shift values in the case of structures 1 and 2, which have a 3- and a 4-membered ring, respectively. The coupling constants, on the other hand, are only compatible with a 6-membered ring in a chair conformation, which leaves structure 4 as correct solution. Vicinal coupling constants of ca. 10 Hz as found for $J_{2,3}$ and $J_{3,4}$ are only observed for the diaxial position of the coupling partners. Axial-equatorial and diequatorial coupling constants are around 4 Hz. The methine proton at δ 4.12 appears at a field higher by ca. 1 ppm than the others so that this CH group must be bonded to an ether oxygen instead of an ester.

Thus, we have derived both the configuration and conformation of the compound as α-D(+)-glucose pentaacetate:

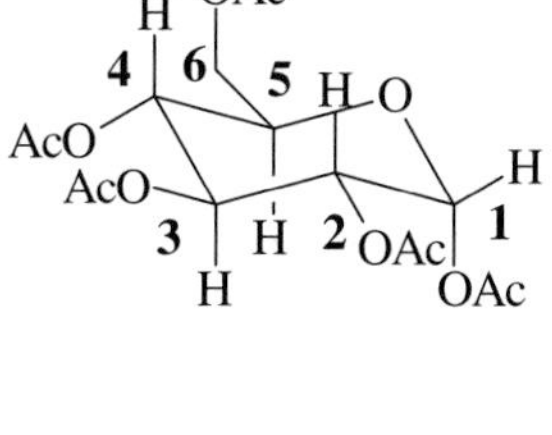

3.16.3 Comments

3.16.3.1 Mass Spectrum

The statement that the occurrence of a fragment with a mass difference of 16 u from the molecular ion is unlikely, needs some explanation. Such a difference is chemically *a priori* not unreasonable because methane and oxygen are possible neutral leaving groups. However, the probability of loss of methane without accompanying loss of methyl (Δm 15) is negligible, while loss of oxygen is restricted to a few types of functional groups and is then either accompanied by loss of 17 u (OH) as in some N-oxides, epoxides, and sulfoxides, or by loss of 30 u as in nitro compounds. Since none of these side reactions occur, the original argument is applicable.

3.16.3.2   Infrared Spectrum

The compound exhibits C–H stretching vibrations just above 3000 cm^{-1}, seemingly indicating hydrogen atoms bonded to sp^2-hybridized carbon atoms or on a three-membered ring. However, in the present case, the solvent absorption mimicks a high frequency. Notwithstanding, there are certain structural elements that may exhibit C–H stretching vibration frequencies in this region, e.g., hydrogen atoms bonded to carbon substituted with halogens.

3.16.3.3 ^{1}H and ^{13}C NMR Spectra

Despite of the small ^{1}H and ^{13}C chemical shift differences, all O-acetyl groups can be assigned on the basis of the HMBC measurement (Fig. 3.16.8). The carbonyl carbon atoms are identified by their cross peaks (Fig. 3.16.8, bottom) originating from the coupling over three bonds to the sugar methine and methylene protons. Also, the CH$_3$ proton signals can easily be assigned from the C=O/CH$_3$ connectivities (see expanded section of the HMBC map).

3.16.3.4   Presentation of NMR Data (500 resp. 125 MHz, CDCl$_3$, δ)

Assignment	^{1}H (J)	^{13}C	Selected HMBC responses (^{13}C partners)
1	6.33, d (3.8 Hz)	89.0	C-3, C-5, C=O (1)
2	5.10, dd (10.0, 3.8 Hz)	69.1	C-3, C=O (2)
3	5.47, t (10.0 Hz)	69.7	C-2, C-4, C=O (3)
4	5.14, t (10.0 Hz)	67.8	C-3, C-5, C=O (4)
5	4.12, ddd (10.0, 4.4, 2.3 Hz)	69.7	–
6a	4.27, dd (12.9, 4.4 Hz)	61.4	C=O (6)
6b	4.10, dd (12.9, 2.3 Hz)	–	–
CH$_3$ (1)	2.18, s	20.8	C=O (1)
CH$_3$ (2)	2.02, s	20.3	C=O (2)
CH$_3$ (3)	2.03, s	20.5	C=O (3)
CH$_3$ (4)	2.04, s	20.4	C=O (4)
CH$_3$ (6)	2.10, s	20.6	C=O (6)
C=O (1)	–	168.7	–
C=O (2)	–	169.6	–
C=O (3)	–	170.1	–
C=O (4)	–	169.3	–
C=O (6)	–	170.5	–

3.17 Problem 17

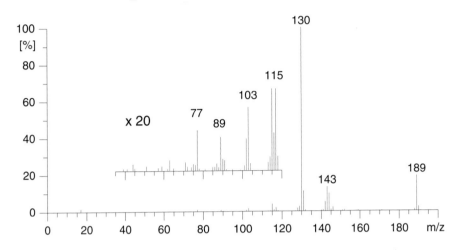

Fig. 3.17.1: Mass spectrum, EI, 70 eV

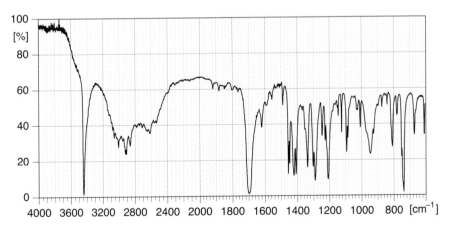

Fig. 3.17.2: IR spectrum, KBr pellet

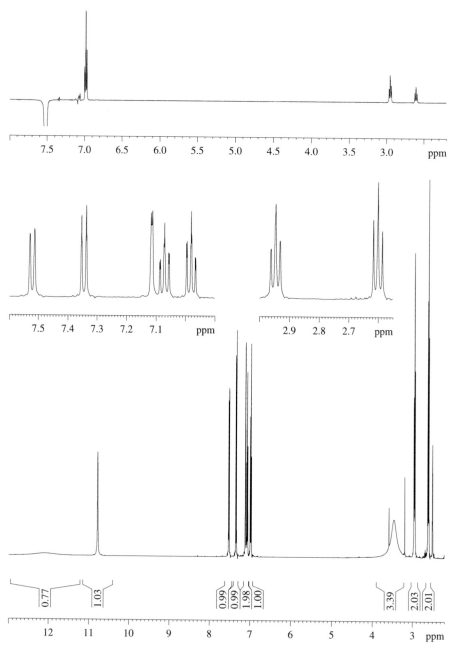

Fig. 3.17.3: ^{1}H NMR spectra, 500 MHz, solvent DMSO-d$_6$. Top: 1D-NOESY spectrum obtained by irradiating the signal at δ 7.52; bottom: conventional spectrum

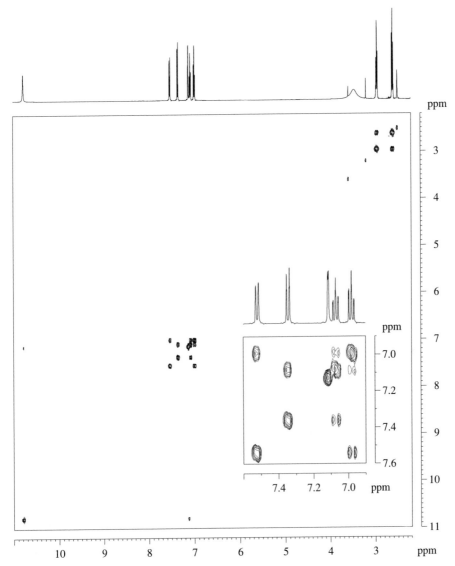

Fig. 3.17.4: ^{1}H,^{1}H COSY spectrum, 500 MHz, solvent DMSO-d$_6$

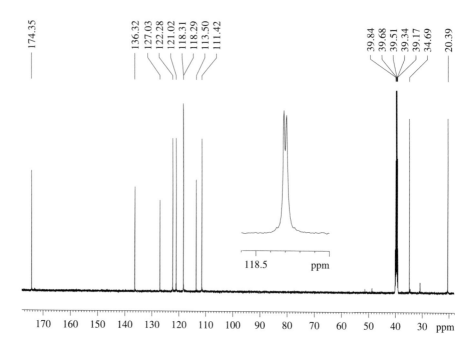

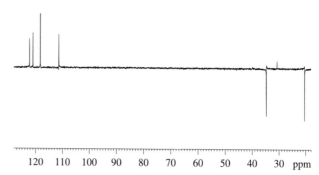

Fig. 3.17.5: ^{13}C NMR spectra, 125 MHz, solvent DMSO-d$_6$. Top: proton-decoupled; bottom: DEPT135

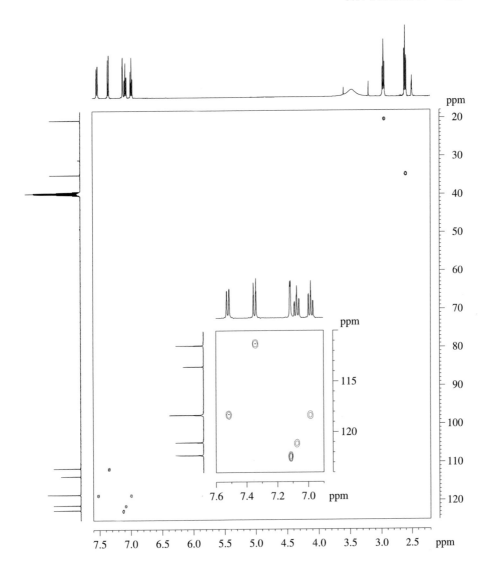

Fig. 3.17.6: $^{13}C,^1H$ HSQC spectrum, solvent DMSO-d$_6$

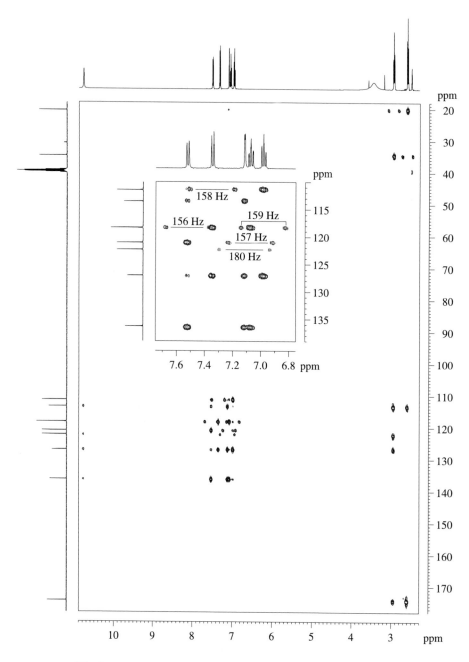

Fig. 3.17.7: $^{13}C,^{1}H$ HMBC spectrum, solvent DMSO-d_6

3.17.1 Elemental Composition and Structural Features

The mass spectrum (Fig. 3.17.1) ends with the peak at m/z 189 that can be attributed to the molecular ion because of the reasonable the mass differences with respect to all fragments. The odd number of M_r indicates the presence of an odd number of nitrogen atoms in the molecule.

The IR spectrum (Fig. 3.17.2) shows a relatively sharp, intense band at 3450 cm^{-1}, which most probably corresponds to an N–H stretching vibration. An NH$_2$ group would have at least two comparable bands, while an OH signal would either show a higher frequency (for the free form) or would be much broader (for an H-bonded form). The broad band at 3300–2400 cm^{-1} can arise from an acid or an ammonium group. We further identify a C=O group (stretching frequency at 1700 cm^{-1}).

The integration of the ^{1}H NMR spectrum taken in DMSO-d$_6$ (Fig. 3.17.3, bottom) leads to intensity ratios of ca. 1:1:1:1:2:1:3.4:2:2 (from left). The broad signal at δ 3.4 must originate from water, practically always present in this hygroscopic solvent, and that at δ 2.5 from the partially nondeuterated DMSO. Two weak signals at δ 3.2 and 3.6 indicate traces of impurities. The remaining signals correspond to 11 H atoms. The two triplets at δ 2.94 and 2.60 indicate the presence of a CH$_2$–CH$_2$ group. According to the COSY spectrum (Fig. 3.17.4), the four peaks at δ 7.52, 7.34, 7.07, and 6.98 with coupling constants of ca. 8 Hz and further small couplings belong to one and the same spin system and can be assigned to an *o*-disubstituted benzene ring. As indicated by the HSQC spectrum (Fig. 3.17.6), there is one more sp^2-hybridized CH group (δ 7.11 and 122.3) and the two remaining protons at ca. δ 12.1 and 10.78 are not bonded to C. The former matches that of a COOH group and the latter must be due to the NH already identified from the IR spectrum. The COSY spectrum shows that the small splitting of the isolated =CH proton comes from a coupling with the NH proton. Because of the bigger line width of the NH signal, its corresponding splitting cannot be seen. The ^{13}C NMR spectra (Fig. 3.17.5) show the presence of 2 CH$_2$, 5 =CH, 3 =C, and one C=O, corresponding to a total of 11 carbon atoms. The elements found so far add up to the molecular formula of C$_{11}$H$_{11}$NO$_2$, which also fits the molecular mass.

3.17.2 Structure Assembly

Check with Assemble 2.1

The structural elements found are summarized in the input window of Assemble 2.1 (Fig. 3.17.8). The atom tags of the CH$_2$CH$_2$ group (<Vh22>) forbid the presence of further vicinal hydrogen atoms.

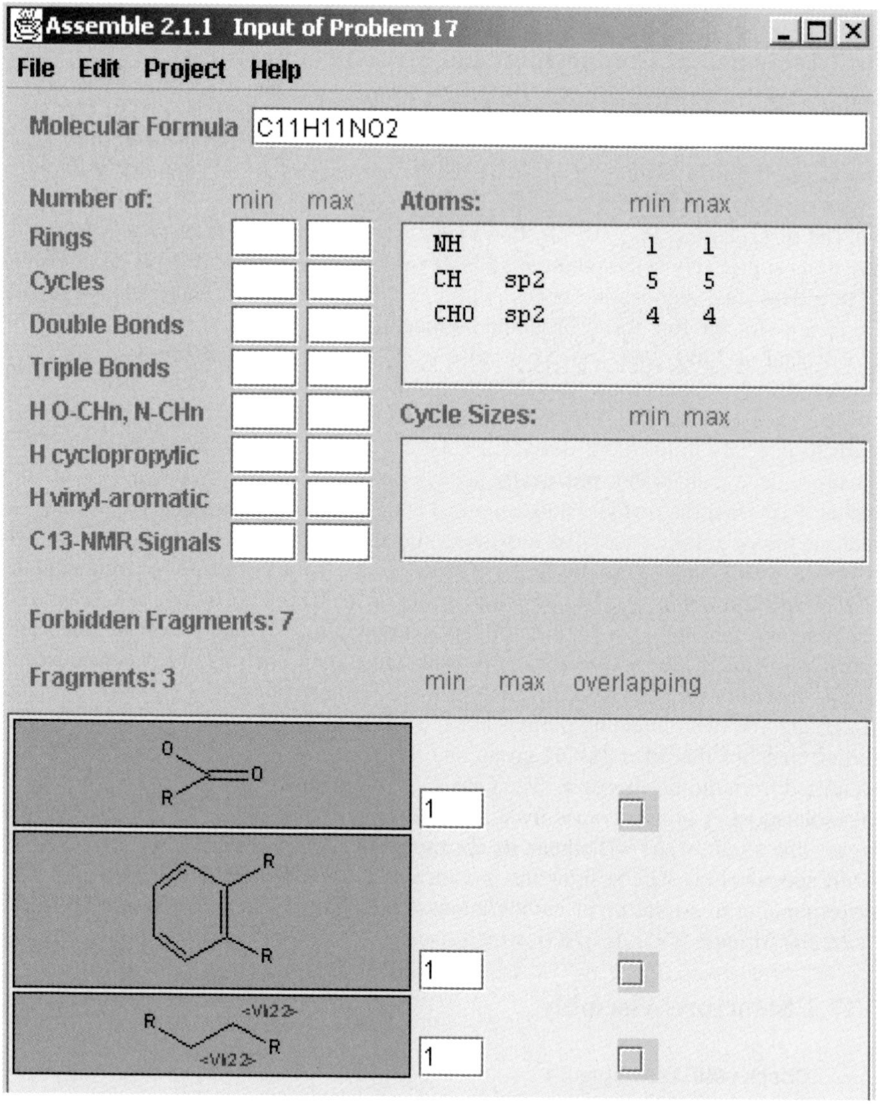

Fig. 3.17.8: Input to Assemble 2.1

With this input, Assemble 2.1 generates 36 constitutional isomers. Their inspection shows that some of them contain a monosubstituted benzene ring. This does not contradict the input since R stands for a free valence that can be occupied by any atom including H. The vicinal hydrogen atom tag can be used on both terminal =CH groups of the CH=CH–CH=CH moiety to prevent the generation of monosubstituted benzenes. Furthermore, the base peak in the mass spectrum at $M^{+\cdot}$ − 59 (loss of CH_2COOH) suggests that the first and third fragments in Fig. 3.17.8 can be combined to form a CH_2CH_2COOH group. With these modifications of the input, Assemble 2.1 generates the three structures shown in Fig. 3.17.9.

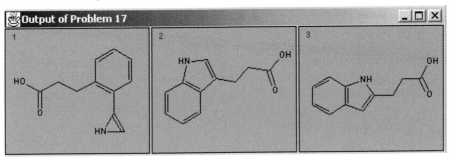

Fig. 3.17.9: Output of Assemble 2.1 after modifying the input

On the basis of the HMBC spectrum (Fig. 3.17.7), which shows a cross peak at δ 2.94/122.3 relating one CH$_2$ to the isolated =CH, structure 1 can be excluded since it would require these groups to be separated by 5 bonds. In the 1D NOESY spectrum (Fig. 3.17.3, top), the irradiation of the doublet at δ 7.52 results in signals for the protons of both CH$_2$ groups. Only isomer 2 fulfills the requested steric proximity.

3.17.3 Comments

3.17.3.1 Infrared Spectrum

The IR spectrum is recorded as a KBr pellet. Since KBr is hygroscopic, a broad water absorbance is usually observed in the range of 3600–3300 cm^{-1}. In the present case, this band is not seen because of the strong signals of the NH and COOH vibrations. Another feature often observed with KBr pellets is the asymmetry of the bands. They show a steep slope at the lower wavelength and a tailing on the other side. Such distortions (called Christiansen effect) are the consequence of strong differences in the refractive indices of KBr and the dispersed compound. They can be reduced by finely grinding the sample.

3.17.3.2 ^{1}H and ^{13}C NMR Spectra

On the basis of the HMBC spectrum (see also Section 3.11.3.3), it is possible to assign all ^{1}H and ^{13}C signals. In this spectrum, we also observe the one-bond couplings, $^1J_{C,H}$. The large value of $^1J_{C,H}$ = 180 Hz of the carbon at δ 122.3 indicates that it is attached to the electronegative N. The cross peak of H-2 with the methylene carbon at δ 20.4 allows the assignment of CH$_2$-8. The correlations of the CH$_2$-9 protons (δ 2.60) give the assignment of the quaternary C-3 and provide a further indication for the attachment of the carboxyl group to the CH$_2$CH$_2$ moiety, as already deduced from the mass spectrum. The correlation of CH$_2$-8 with the quaternary carbon atom at δ 127.0, on the other hand, provides the assignment of C-3a. Considering the fact that in aromatic systems, $^2J_{C,H} < {}^3J_{C,H}$, the HMBC correlations of C-3a assign H-5 and H-7, while the responses of C-7a determine H-4 and H-6.

3.17.3.3   Presentation of NMR Data (500 resp. 125 MHz, DMSO-d$_6$, δ)

Assignment	^{1}H (J)	^{13}C	HMBC responses (^{13}C partners)
1	10.78	–	C-2, C-3, C-3a, C-7a
2	7.11	122.3	C-3, C-3a, C-7a
3	–	113.5	–
3a	–	127.0	–
4	7.52, d (8.0 Hz)	118.29	C-3, C-3a, C-6, C-7a
5	6.98, td (8.0, 1.5 Hz)	118.31	C-3a, C-7
6	7.07, td (8.0, 1.5 Hz)	121.0	C-4, C-7a
7	7.34, d (8.0 Hz)	111.4	C-3a, C-5
7a	–	136.2	–
8	2.94, t (7.0 Hz)	20.4	C-2, C-3, C-3a, C-9, C-10
9	2.60, t (7.0 Hz)	37.4	C-3, C-8, C-10
10	–	174.4	–
OH	12.09	–	–

3.18 Problem 18

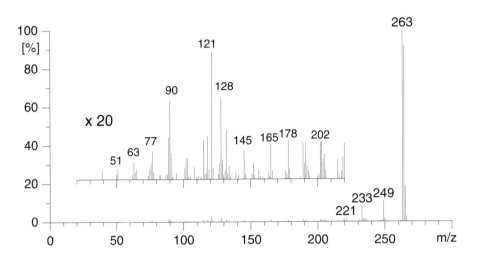

Fig. 3.18.1: Mass spectrum, EI, 70 eV

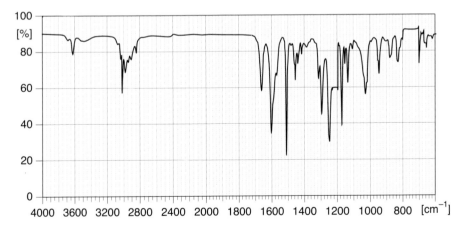

Fig. 3.18.2: IR spectrum, solvent CHCl₃, cell thickness 0.2 mm

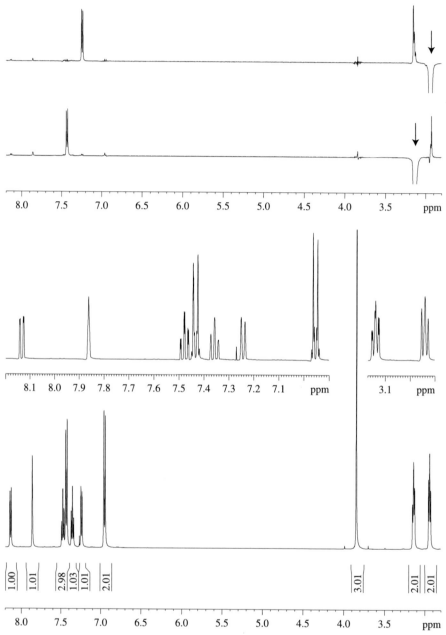

Fig. 3.18.3: ^{1}H NMR spectra, 500 MHz, solvent CDCl$_3$. Bottom: conventional spectrum; top two traces: 1D NOESY spectra obtained by irradiating the signals marked with an arrow

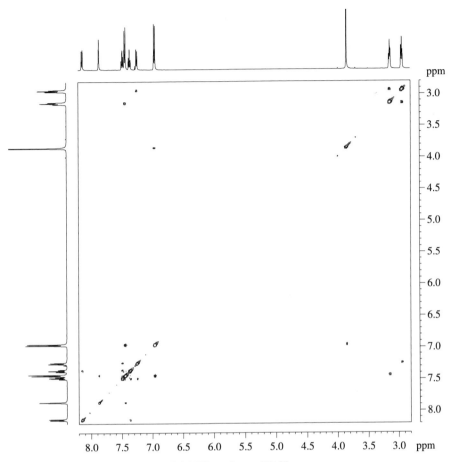

Fig. 3.18.4: NOESY spectrum, 500 MHz, solvent CDCl$_3$

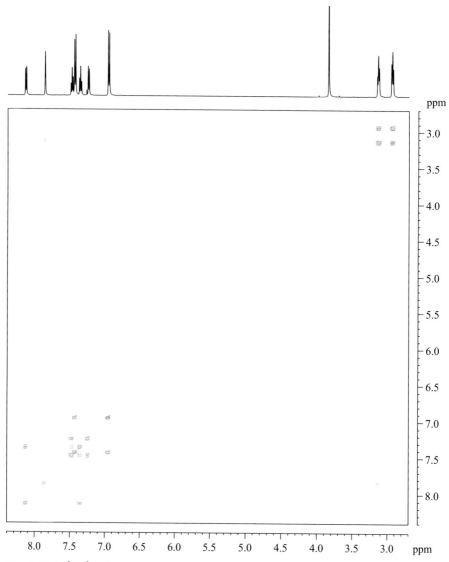

Fig. 3.18.5: ^{1}H,^{1}H COSY spectrum, 500 MHz, solvent CDCl$_3$

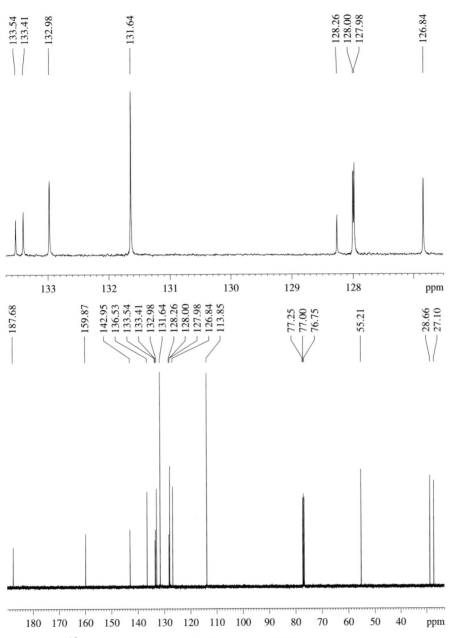

133.54
133.41
132.98
131.64
128.26
128.00
127.98
126.84

| 133 | 132 | 131 | 130 | 129 | 128 | ppm |

187.68
159.87
142.95
136.53
133.54
133.41
132.98
131.64
128.26
128.00
127.98
126.84
113.85
77.25
77.00
76.75
55.21
28.66
27.10

| 180 | 170 | 160 | 150 | 140 | 130 | 120 | 110 | 100 | 90 | 80 | 70 | 60 | 50 | 40 | ppm |

Fig. 3.18.6: ^{13}C NMR spectrum, 125 MHz, solvent CDCl$_3$

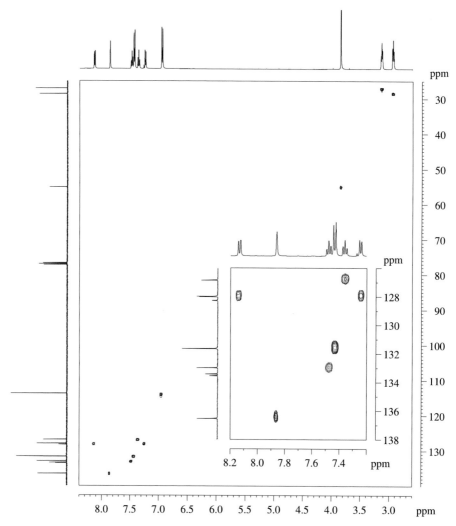

Fig. 3.18.7: $^{13}C,^{1}H$ HSQC spectrum, solvent CDCl$_3$

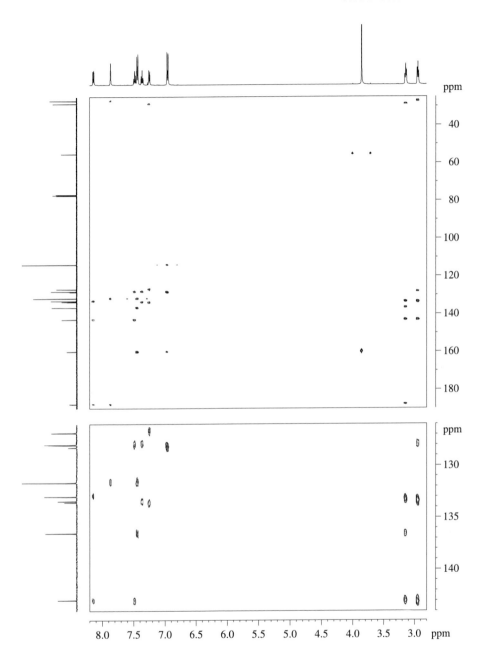

Fig. 3.18.8: ^{13}C,^{1}H HMBC spectrum, solvent CDCl$_3$

3.18.1 Elemental Composition and Structural Features

The mass spectrum (Fig. 3.18.1) ends with intensive peaks at m/z 263 and 264. We assume that the latter corresponds to the molecular ion because the fragment at m/z 249 must originate from m/z 264, the loss of 14 u being observed in very special cases only (cf. Problem 8). Both the intensity distribution and the series of unsaturated hydrocarbon fragments in the lower mass range (m/z 39, 51, 63, 77, 90, …) suggest an aromatic character of the molecule.

The IR spectrum (Fig. 3.18.2) shows CH vibrations above and below 3000 cm^{-1} with a peak at 2840 cm^{-1} at the lower end, which indicates an CH$_2$ or CH$_3$ group attached to O or N. The broad intense bands at 1665 and 1605 cm^{-1} are not easy to assign at this stage. The signals at 3680 and 3620 cm^{-1} originate from water as impurity in the solvent.

The integration of the ^{1}H NMR spectrum (Fig. 3.18.3, bottom) leads to intensity ratios of 1:1:3:1:1:2:3:2:2 (from left) corresponding to a total of 16 H atoms in the molecule. The two triplet-like signals at δ 3.14 and 2.94 originate from a CH$_2$CH$_2$ group. The singlet of 3H at δ 3.85 with the corresponding ^{13}C shift of δ 55.2 (Fig. 3.18.6), as shown by the HSQC spectrum (Fig. 3.18.7), must belong to a CH$_3$O group. The remaining 9 H atoms are found in the chemical shift range of alkene/aromatic protons. At δ 6.96, we find a higher-order multiplet of doublet character with an integral of 2H. A closer inspection reveals its symmetric partner at δ 7.43. The corresponding ^{13}C NMR signals at δ 113.9 and 131.6 have intensities about twice as high as the others. These signals indicate a *p*-disubstituted benzene ring having an *AA'XX'* spin system in the ^{1}H NMR spectrum. One of the substituents must account for the strong shielding of one pair of protons and carbon atoms (δ 6.96/113.9) as compared with unsubstituted benzene (δ 7.26/128.5). The CH$_3$O group is a likely candidate for such a substituent. Four of the five remaining protons, appearing as two doublets and two triplets with some additional small couplings, constitute a coupling network (CH=CH–CH=CH) also supported by the COSY spectrum (Fig. 3.18.5). Because of the values of the vicinal coupling constants of 8.5 Hz, they can be assigned to an *o*-disubstituted benzene ring.

Based on this interpretation, the 16 signals in the ^{13}C NMR spectrum correspond to 18 carbon atoms. According to their chemical shifts, three of them are *sp*^{3}-hybridized, one (δ 187.7) must belong to a C=X (most probably C=O) group so that in addition to the two benzene rings there is a further C=CH double bond in the molecule.

The elements found so far lead to the formula $C_{18}H_{16}O_2$ having the assumed molecular mass of 264 u. This elemental composition corresponds to 11 double bond equivalents. Two benzene rings, the carbonyl group, and the C=CH double bond account for 10 of them, therefore, an additional ring must be present.

So far, the complete information from the COSY and HSQC spectra has been used and the signal patterns in the ^{1}H NMR spectrum are fully understood. Additional connectivity information is available from the HMBC spectrum (Fig. 3.18.8). However, since not all ^{13}C-^{1}H couplings over two and three bonds lead to a cross peak it would be cumbersome and time-consuming to enlarge the substructures already found or to connect two of them. In this situation, it is much more straightforward to enter the available information to Assemble 2.1 and eliminate incompatible candidate structures on the basis of spectral information.

3.18.2 Structure Assembly

Check with Assemble 2.1

The structural elements found are summarized in the input window of Assemble 2.1 (Fig. 3.18.9). The atom tags of the CH_2CH_2 group forbid the presence of further vicinal hydrogen atoms and that of the carbonyl group excludes an aldehyde. In the upper part of the input window, cyclopropyl rings and aliphatic CH groups are excluded, and it is required that all nonprotonated carbon atoms are sp^2-hybridized. Finally, the symmetry constraint requires 16 signals for the 18 carbon atoms. Since this is implied by the first substructure in Fig. 3.18.9, no further symmetry can occur in the molecule.

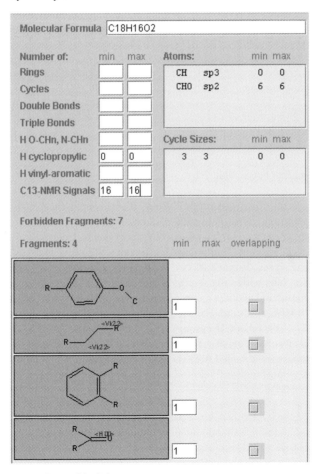

Fig. 3.18.9: Input to Assemble 2.1

With this input, Assemble 2.1 generates 13 structures (see Fig. 3.18.10). The correct solution can be found by interpreting the HMBC (Fig. 3.18.8) and NOESY spectra (Fig. 3.18.3, top traces and Fig. 3.18.4).

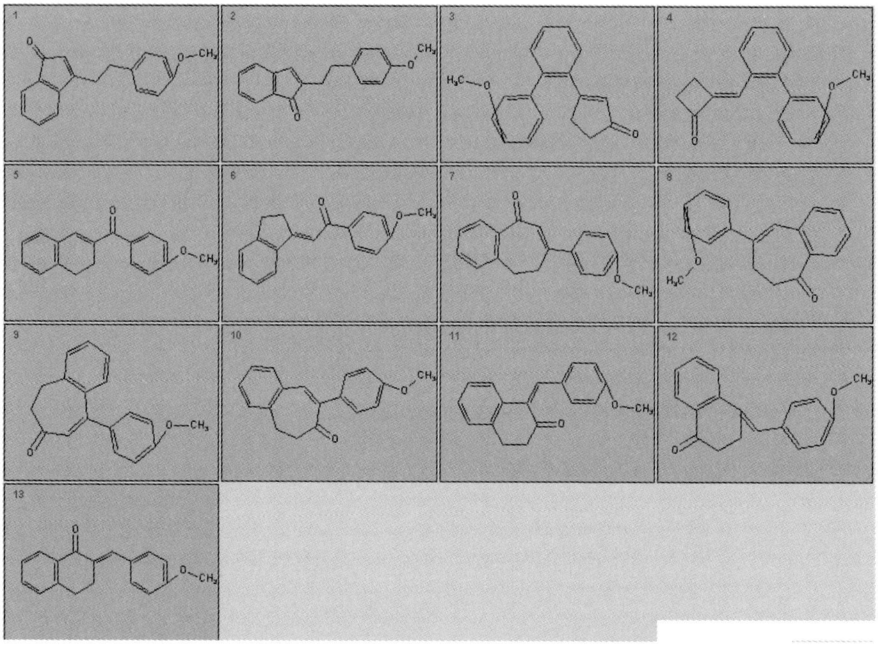

Fig. 3.18.10: Output of Assemble 2.1

In HMBC spectra, cross peaks appear if the ^{13}C-^{1}H coupling constants are within the values selected through the experimental parameters (for all examples in this book: 7 ± (3–4) Hz). These are most often couplings over three bonds, frequently over two bonds, and in exceptional cases over four bonds. The absence of a cross peak cannot be used as an argument because a distance of two or three bonds does not guarantee that the coupling constant falls within the selected range. In the following, two- and three-bond couplings are considered and it is assumed that no coupling over four bonds gives rise to a signal in the HMBC spectrum. For the CH_2 protons at δ 2.94, four cross peaks can be observed at δ 27.1 (the other CH_2 group), 133.5, 143.0 (both =C), and 128.0 (a terminal =CH of the *o*-disubstituted benzene). This implies the following partial structure (the ^{1}H shifts in parentheses are obtained from the HSQC spectrum):

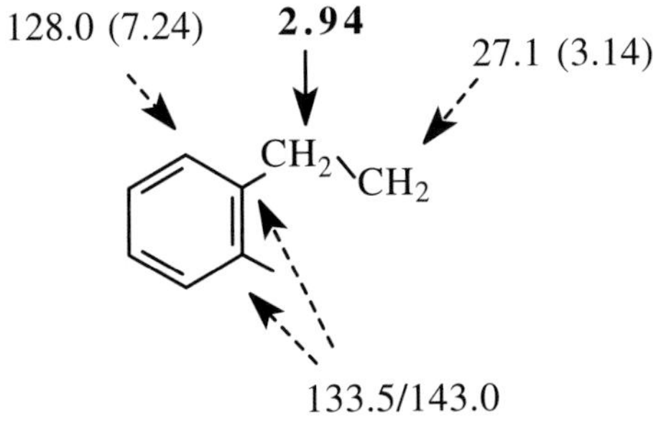

128.0 (7.24)   **2.94**

27.1 (3.14)

CH$_2$ CH$_2$

133.5/143.0

The solutions 1–4, 8, and 12 in Fig. 3.18.10 can be excluded since they are not compatible with this partial structure.

Five HMBC responses of the other CH$_2$ group at δ 3.14 are obtained, namely 187.7 (C=O), 143.0 (=C), 136.5 (=CH), 133.4 (=C), and 28.7 (CH$_2$). These results can be represented by the following partial structure, which rules out solutions 6, 7, 10, and 11:

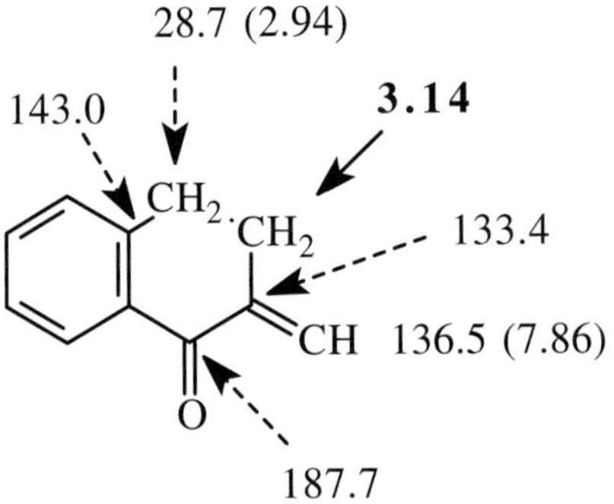

28.7 (2.94)

143.0

3.14

CH$_2$ CH$_2$

133.4

CH 136.5 (7.86)

O

187.7

The first two of the three remaining solutions (5, 9, and 13) can be excluded on the basis of the cross peaks of the H atom at δ 7.86, which appear at δ 187.7 (C=O), 27.1 (CH$_2$), and 131.6. According to the HSQC spectrum, the latter C atoms are directly attached to the protons at δ 7.43, which are part of the *AA'XX'* system:

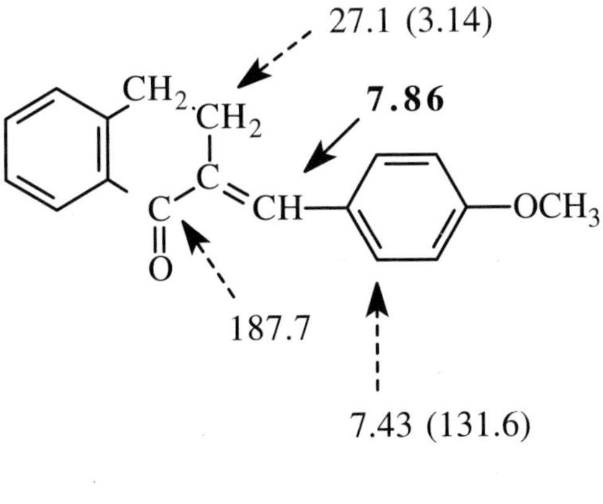

27.1 (3.14)

CH$_2$ CH$_2$

7.86

C

C=CH—⟨ ⟩—OCH$_3$

O

187.7

7.43 (131.6)

The remaining structure 13 in Fig. 3.18.10 can have either *Z* or *E* configuration:

Structure 13 (*E*)   Structure 13 (*Z*)

A decision is possible with the help of the NOESY spectra (Fig. 3.18.3, top traces and Fig. 3.18.4). The appearance of the signals for two protons of the *AA'XX'* system at δ 7.43 upon irradiation of one of the CH_2 groups (δ 3.14) is only possible with structure 13 (*E*). The second 1D NOESY experiment (Fig. 3.18.3, top) shows the steric proximity of the other CH_2 group (δ 2.94) to one proton of the second benzene ring and, thus, further supports the above interpretation of the HMBC spectrum.

3.18.3 Comments

3.18.3.1 Assemble 2.1

Ranking on the basis of NMR chemical shifts is not powerful for isomers generated from rather large substructures since the differences in their shift values are small. In the present case, the correct structure is found as the best one with the ^{13}C NMR ranking and has 1.4 and 4.6 ppm as mean and maximum deviations, respectively. Based on the ^1H NMR spectrum, it has position 4 with respective deviations of 0.27 and 0.65 ppm. However, it would be risky to exclude most of the other structures in Fig. 3.7.10 because only in five cases, the mean and maximum deviations are larger than 5 and 20 ppm, respectively, for ^{13}C NMR and in all cases, they are smaller than 0.35 and 0.90 ppm for ^1H NMR.

3.18.3.2 Infrared Spectrum

Owing to strong delocalization, the frequency of the C=O stretching vibration is very low. The unusual intensities of the bands at 1665 and 1605 cm^{-1} indicate a strong coupling between C=O and C=C vibrations.

3.18.3.3 ^1H and ^{13}C NMR Spectra

An unambiguous assignment of the quaternary carbon signals is possible on the basis of the HMBC spectrum. The two- and three-bond ^1H-^{13}C responses are summarized in the Table of 3.18.3.4. Both the CH_2-3 (δ 3.14) and H-8 (δ 8.13) protons show a cross peak with the signal at δ 143.0, which must correspond to the quaternary C-4a carbon atom. Over three bonds, H-7 and also H-5 point to the signal at δ 133.5, which is assigned to C-8a. In the *p*-disubstituted phenyl ring, the methoxy protons point to C-4', and the H-3',5' protons to C-1'.

As mentioned previously, sometimes also the one-bond $J_{C,H}$ correlations break through in the HMBC spectrum. This can be observed here with the CH$_3$O– (δ 3.85/55.2), H-2',6'/C-2',6' (δ 7.43/131.6), and H-3',5'/C-3',5' (δ 6.96/113.9) signals. Note that for these aromatic CH units the real cross peaks are also observed in that H-2' and the isochronous H-6' mark out over three bonds the carbon atoms C-6' and C-2', respectively.

3.18.3.4 Presentation of NMR Data (500 resp. 125 MHz, CDCl$_3$, δ)

Assignment	^{1}H (*J*)	^{13}C	HMBC responses (^{13}C partners)	NOE responses (^{1}H)
1	–	187.7	–	–
2	–	133.4	–	–
3	3.14, m	27.1	C-1, C-2, C-4, C-4a, C-9	4, 2'/6'
4	2.94, m	28.7	C-3, C-4a, C-5, C-8a	3, 5
4a	–	143.0	–	–
5	7.24, d (8.5 Hz)	128.0	C-4, C-7, C-8a	4, 6
6	7.48, t (8.5Hz)	133.0	C-4a, C-8	5, 7
7	7.36, t (8.5 Hz)	126.8	C-5, C-8a	6, 8
8	8.13, d (8.5 Hz)	128.0	C-1, C-4a, C-6	7
8a	–	133.5	–	–
9	7.86, s	136.5	C-1, C-3, C-2'/6'	2'/6'
1'	–	128.3	–	–
2', 6'	7.43, m	131.6	C-9, C-2'/6'	3, 9, 3'/5'
3', 5'	6.96, m	113.9	C-1', C-3', C-4', C-5'	2'/6', CH$_3$O
4'	–	159.9	–	–
CH$_3$O	3.85, s	55.2	C-4'	3'/5'

4 Additional Remarks

4.1 Mass Spectrometry

4.1.1 Presentation of Data

Mass spectra were recorded on a VG-Tribid or a Hewlett Packard 5970 mass spectrometer and are presented as linear line graphs with mass-numbered intensity maxima, normalized to the strongest peak (base peak) of the spectrum. This is the most convenient summary of the data for a quick survey and makes characteristic features like intensity distribution, ion series, or isotope pattern easily perceptible, thereby outweighing the disadvantages against tabulated spectra of less accurate intensity specification and loss of weak signals.

4.1.2 Degree of Unsaturation, Calculation of the Number of Double Bond Equivalents

The elemental composition of a saturated hydrocarbon is C_nH_{2n+2}. Incorporation of k double bonds or rings is tantamount to a removal of $2k$ hydrogen atoms from that formula. A triple bond is equivalent to two double bonds. Halogen atoms can replace hydrogen. Thus, F, Cl, Br, and I are hydrogen equivalents; their number is added to the number of hydrogen atoms when calculating the degree of unsaturation. In the same sense, Si, Ge, Sn, and Pb are carbon equivalents; their number is added to the number of carbon atoms. Insertion of O (or of the oxygen equivalents S, Se, and Te in their divalent form) does not alter the carbon/hydrogen ratio in saturated systems so that they can be neglected in these calculations. Insertion of m trivalent N atoms (or of the nitrogen equivalents P, As, Sb, or Bi) requires addition of m hydrogen atoms to maintain saturation so as to obtain the general formula for saturated systems, $C_nH_{2n+2+m}N_m$, where N can be replaced by any of the nitrogen equivalents. For compounds containing atoms other than the aforementioned, the degree of unsaturation is preferably evaluated by applying the general equation 4.2.

The number of double bond equivalents (or degree of unsaturation), F, corresponds to the difference between the number of required hydrogen atoms for complete saturation, $(2n + 2 + m)$, and the actual number, x, (including hydrogen equivalents) divided by two:

$$F = \frac{(2n + 2 + m) - x}{2} \tag{4.1}$$

where n represents the number of carbon atoms plus carbon equivalents, m the number of nitrogen atoms plus nitrogen equivalents, and x the number of hydrogen atoms plus hydrogen equivalents.

In a simplified approach, which suffices for most common cases, i.e., for compounds containing only C, H, O, trivalent N, divalent S, and the halogens, one can reduce the elemental composition formula to a hydrocarbon equivalent and calculate the degree of unsaturation according to the following steps:

1. omit oxygen and sulfur,
2. replace all halogens by hydrogens,
3. replace all nitrogens by CH groups,
4. compare the resulting hydrocarbon composition C_nH_x with the composition of a saturated hydrocarbon C_nH_{2n+2}: The number of unsaturations is given by half the number of missing hydrogens.

In a comprehensive approach of general applicability, the number of double bond equivalents is calculated by the following equation:

$$F = \frac{2 + \sum_i n_i(v_i - 2)}{2} \tag{4.2}$$

where n_i is the number of the element i having a formal valence of v_i.

Even-electron fragment ions in the mass spectrum (odd-mass ionic species if they contain no or an even number of nitrogen atoms) are species with one open valence. A saturated system of this kind, therefore, contains one hydrogen less than specified above. This should be taken into account when calculating the degree of unsaturation for fragments.

If the nominal mass of the molecular ion is even, the number of hydrogen atoms is even as well unless an odd number of hydrogen equivalents is present, and *vice versa*.

4.1.3   General Information from Mass Spectra

The intensity distribution of the signals in mass spectra reflects to some extent stability features of the investigated structures. Concentration of the overall ion yield in the molecular ion region indicates a compact molecular arrangement as in purely aromatic compounds, largely conjugated or unsaturated polycyclic systems, and the like, while in saturated aliphatic compounds, the low mass range carries most of the total ion yield. Stable entities substituted by a few easily removable residues produce a few significant intensity maxima as common sense would suggest.

Sequences of intensity maxima (ion series) in the lower mass range and their respective m/z values are indicators of structural type and degree of saturation, which constitute valuable initial information for interpretation. The series $15 + (14)n$ (m/z 29, 43, 57, 71 ..., "alkyl series") of elemental composition C_nH_{2n+1} or $C_nH_{2n-1}O$ is typical of saturated aliphatic hydrocarbons or ketone and aldehyde compounds or residues, while the series $13 + (14)n$ (m/z 27, 41, 55, 69 ..., "alkenyl series") indicates one double bond equivalent as in alkenes, cycloalkanes, and cycloalkanones or monofunctionalized compounds, which easily eliminate a neutral molecule (e.g., water from alcohols). Aromatic hydrocarbon residues result from degradation in the highly unsaturated "aromatic series" (m/z 39, 51 ± 1, 64 ± 1, 78 ± 1, 91 ...). Singly bonded oxygen in saturated systems gives rise to the "oxygen series" (m/z 31, 45, 59, 73 ...), nitrogen in aliphatic saturated compounds to the "nitrogen series" (m/z 30, 44, 58, 72 ...), sulfur in saturated residues to the "sulfur or 2 oxygen series" (m/z 47, 61, 75, 89 ...). Polycyclic saturated systems cause a gradual switching of sequences of maxima into more and more unsaturated series along the way up the mass scale. In general, the intensity distribution within such series is steadily rising or falling. Striking intensity jumps (positive or negative) of individual members within a series are always structurally significant and should be interpreted. If isolated intensity maxima are observed in the lower mass range, it is usually rewarding to determine which ion series they belong to and consider their possible elemental composition. Even-mass maxima within an uneven series, and *vice versa*, are always diagnostically important features too.

4.1.4 Evidence for Elemental Composition from Isotope Peak Intensities

Most of the elements constituting organic molecules naturally occur as mixtures of different isotopic species. Their natural relative abundance is not constant in a strict quantitative sense but constant enough to be characteristic of the individual element within the accuracy limits of intensity measurements in normal qualitative analysis by mass spectrometry (standard deviations significantly < 1% of the mean).

Since mass differences between particular isotopes are very close to full mass units or multiples thereof, ions of a given elemental composition will always yield more than one signal in the mass spectrum. Species containing heavy isotopes are separated in the analyzer and give rise to isotope peaks whose intensity is related to the type and number of individual atoms in the molecular formula by their respective natural isotope abundances. Among the most common elements in organic chemistry, F, P, and I are monoisotopic, and the natural abundance of heavy isotopes of H, N, and O is too low to produce isotope peaks of sufficient intensity. In contrast, the natural abundance of ^{13}C in C is significant enough, each carbon atom contributing about 1.1% to the isotope peak intensity at the next integer mass value of singly charged ions. Consequently, the intensity of a signal in % of that at the previous integer mass value divided by 1.1 indicates the upper limit for the number of carbon atoms that can be present in the respective ionic species. Owing to protonation, which is fairly common, or due to contributions from other elements with appreciable natural abundance of heavy isotopes, (e.g., from Si: 5.1% abundance of ^{29}Si relative to ^{28}Si), the intensity of the first isotope peak can be higher, but never lower, than required by the number of carbon atoms present.

In the second isotope peak, the contribution due to the common elements C, H, N, O (i.e., the probability of an ion containing two ^{13}C, ^{2}H, ^{15}N, or ^{17}O, or one ^{18}O) can be assumed to be approximately 10% of the intensity of the first ^{13}C isotope peak. This is a very rough approximation indeed, but sufficient for purposes of qualitative analysis (deviation less than 1% relative to the intensity of the $M^{+\cdot}$ signal). If the first isotope peak contains significant amounts of protonated species without ^{13}C, the corresponding ^{13}C isotope intensity must be taken into account in the second isotope peak. Intensities of second isotope peaks in excess of these values indicate elements with higher natural abundances of heavy isotopes two mass units apart, such as Si (^{30}Si of 3.3% abundance relative to ^{28}Si), S (^{34}S of 4.5% abundance relative to ^{32}S), Cl (^{37}Cl of 32.0% abundance relative to ^{35}Cl), Br (^{81}Br of 97.3% abundance relative to ^{79}Br), or elements occurring less frequently in organic compounds. The presence of several such elements gives rise to characteristic peak clusters, the intensity distribution of which can be calculated from natural abundances.

4.1.5   Evidence for Elemental Composition in Low Resolution Mass Spectra

Structurally nonspecific information concerning the elemental composition or presence or absence of specific hetero atoms can be drawn not only from characteristic isotope peak intensities but also from characteristic mass values of fragment peaks or of mass differences between molecular ion and fragments.

4.1.6   High Resolution Data

If elemental composition assignments are based on accurately measured mass values, two limiting conditions must always be kept in mind:
1. The correlation between accurate mass and elemental composition is significant only within the limits of resolving power achieved in the measurement and always provided that the assumption of a uniform elemental composition of the measured ion beam is correct. In an unresolved multiplet, the measured mass value will represent the weighted average of all components contributing to the respective peak and its correlation with one specific elemental composition is bound to fail if the standard deviation of mass measurement is smaller than the mass difference between individual compound masses and their weighted average.
2. Irrespective of the accuracy of measurement, the correlation between accurate mass value and elemental composition is ambiguous at higher masses if many hetero atoms must be allowed for. There is always more than one possible combination of elements yielding the same accurate mass, the number increasing with increasing number and type of atoms to be considered. An appropriate choice among these can only be made by controlling the respective isotope peak intensities and masses, by consistency checks over wide ranges of the spectrum, and/or by including additional information (e.g., other spectroscopic data or results of combustion analysis) as selective arguments.

4.1.7   Impurities in Mass Spectra

No sample ever is absolutely pure. The most common impurities are traces of solvents, or phthalates from plasticizers in commercial polymers or from pump fluids, or traces of homologous compounds usually contained in reagents as well as natural products. Of course, other impurities may often be present as well and complicate the interpretation. Their spectra overlay those of the main compound and give rise to ambiguities unless they are identified as not belonging to the subject of analysis.

The nature of sample admission techniques and the high sensitivity inherent in mass spectrometry can lead to spectra that do not reflect the relative amount of impurity. If the sample is not completely evaporated and admitted from a reservoir, fractionation effects may occur, the extent of which depends on the difference in vapor pressure between compound and impurity at the given temperature. In individual scans, enrichments by factors of thousands may result and preclude even a qualitative estimation of the amount of impurity if differences in volatility are large. Consequently, there is usually little point in trying to verify through other spectroscopic data the presumed impurities gathered from mass spectra, unless comparable volatilities (as in homologous compounds) can reasonably be assumed. Signals due to impurities in the molecular ion region can interfere with the correct determination of the molecular mass because they may mimic chemically unreasonable mass differences and thereby suggest higher values than are actually involved. If ambiguity cannot be eliminated by considering mass differences and establishing two independent degradation series, the spectrum needs to be rerun and fractionation verified by comparing mass spectra obtained at different temperatures.

Identification of the correct molecular mass or even the correct elemental composition is, by the same token, not a reliable indication for sample purity because impurities may have been fractionated off before the specific spectrum was registered, or they may be of lower mass, or be insufficiently volatile under the given experimental conditions and thereby be impossible to detect, or they may simply be isomers.

4.2 Infrared Spectroscopy

4.2.1 Presentation of Data

The IR spectra were recorded on a Perkin-Elmer 1600 series Fourier transform spectrophotometer and are presented as plots of transmittance versus frequency. The frequency scale is inverted, the frequencies increasing from right to left. Frequency values are given as wave numbers, v, in cm^{-1}. There is a change of scale at $v = 2000$ cm^{-1}. The scale to the right of this value is expanded by a factor of 2 relative to that to the left. The matrix used for recording the IR spectra is given at the bottom.

4.2.2 Prediction of Infrared Stretching Frequencies

A rough prediction of IR stretching frequencies can be made using the harmonic oscillator model. The absorption frequency of the stretching mode of a two-atom oscillator depends primarily on the mass of the two atoms involved and on the force constant of the bond between them. It may be estimated according to the following equation:

$$v_{st}[cm^{-1}] = 1303 \sqrt{k\left(\frac{1}{m_1} + \frac{1}{m_2}\right)} \qquad (4.3)$$

where m_1 and m_2 are the relative atomic masses; k is a factor characterizing the type of bond and can assume values of 5, 10, and 15 for single, double, and triple bonds, respectively. Due to the many simplifying assumptions (e.g., the harmonic oscillator is fully independent of the rest of the molecule), predicted values provide an order of magnitude only. The equation is, however, useful for predicting isotope shifts, as exemplified in the following.

Predicted for C–H st:

$$v_{C-H} = 1303 \sqrt{5\left(\frac{1}{12} + \frac{1}{1}\right)} = 3033\,cm^{-1} \qquad (4.4)$$

Prediction for C–D st:

$$v_{C-D} = 1303 \sqrt{5\left(\frac{1}{12} + \frac{1}{2}\right)} = 2225\,cm^{-1} \qquad (4.5)$$

Thus, C–D stretching frequencies in deuterated hydrocarbons are expected around 2200 cm^{-1}.

If the stretching vibration frequency, v_o, of the unlabelled compound is known, the respective frequency, v_L, of the labelled compound can be predicted with considerable accuracy by using the following relation:

$$v_L = v_o \sqrt{\frac{\dfrac{1}{m_1} + \dfrac{1}{m'_2}}{\dfrac{1}{m_1} + \dfrac{1}{m_2}}} \qquad (4.6)$$

where m_1, m_2, and m'_2 are the relative atomic masses of the atoms 1 and 2 involved and of the isotope, respectively.

As an example, the D–O stretching frequency in liquid D_2O is calculated from the H–O stretching frequency in H_2O (3490 cm^{-1}) as:

$$v_{D-O} = 3490\,cm^{-1} \sqrt{\frac{\dfrac{1}{16} + \dfrac{1}{2}}{\dfrac{1}{16} + \dfrac{1}{1}}} = 2539\,cm^{-1} \qquad (4.7)$$

which compares very well with the observed value of 2540 cm^{-1}.

4.2.3 Overtones, Combination Bands, and Fermi Resonance

In general, IR absorption bands correspond to transitions from the ground state to the first vibrationally excited state. However, in some cases the vibration quantum number may change by 2. The corresponding absorption bands, the so-called overtones, are found at frequencies equal to approximately twice the frequency of the fundamental band. Furthermore, the absorption of IR radiation may cause the simultaneous excitation of two or even three vibrational modes, giving rise to combination bands. These are found at frequencies corresponding to the sum (the molecule gains energy in both modes) or to the difference (the molecule gains energy in one mode, but loses a smaller amount in the other mode) of the fundamental frequencies involved.

Overtones and combination bands generally exhibit much lower absorption intensities than fundamental vibrations. Nevertheless, they can sometimes be of considerable diagnostic value. For instance, the series of overtones and combination bands observed with benzene derivatives between 2000 and 1600 cm^{-1} are useful for identifying the substitution pattern. Overtones of the out-of-plane deformation vibrations of the hydrogen atoms in terminal methylene groups, found near 1850–1800 cm^{-1}, are helpful for the identification of this structural element.

It may happen that an overtone or combination band accidentally falls very close to a fundamental frequency. If both are of the same symmetry type, the two transitions will interact to give two new transitions, one of higher energy (higher frequency) and one of lower energy (lower frequency) than the original pair. In addition, the total intensity is redistributed between the two new transitions in such a way as to give two bands of similar absorption intensity. This type of interaction is termed Fermi resonance. Thus, a normally weak overtone or combination band may gain enough intensity to become an important absorption band. Some structural types reliably show the phenomenon of Fermi resonance in their spectra. An example are the chlorides of benzoic acids. Here, the carbonyl stretching frequency interacts with the overtone of a band near 875 cm^{-1}, giving rise to two bands in the carbonyl region near 1775 and 1745 cm^{-1}. The two bands, characteristic of aldehydes, which are observed on the low frequency side of the C–H stretching region are most probably also due to Fermi resonance (interaction of the C–H stretching vibration with an overtone of the C–H deformation vibration).

4.2.4 Band Shapes and Intensities in Infrared Spectra

In contrast to spectroscopy in the ultraviolet and visible wavelength region, it is not easily possible to give quantitative measures for absorption band intensities in IR spectroscopy. The reason for this lies primarily in the fact that monochromacity and wavelength resolution of today's instruments are of the same order of magnitude as the bandwidth of IR absorption bands, which strongly affects the band shape. Thus, in qualitative organic analytical chemistry, the absorption intensity is generally characterized by subjective classifications, e.g., weak, medium, strong, and very strong. In the regions where the transmission of the solvent is less than a few %, the light energy reaching the detection system is insufficient for reliable operation. Here, the recording system behaves erratically, no interpretation of the spectra is possible, and the output generally is a horizontal line.

Not all vibrational modes give rise to an IR absorption band. A prerequisite for a vibration to be IR active is that the vibration must cause a change in the dipole moment. For example, the bending and asymmetric stretching modes of CO_2 (which, of course, is a linear molecule) cause the molecule to have a dipole moment except in its equilibrium structure. Hence, both vibrations lead to an IR absorption. In the symmetric stretching vibration, the dipole moment does not change so that there is no IR absorption corresponding to this vibration and the symmetric stretching mode is not IR active. Raman spectroscopy has different selection rules. For a vibration to be Raman active, a change in polarizability of the molecule during the vibration is required. For example, in the symmetric stretching vibration of CO_2, the electrons get farther away from the nuclei as the bond stretches. Thus, they are less strongly attracted to the nuclei and are more polarizable. The inverse holds when the bond shortens. Consequently, the symmetric stretching vibration of CO_2 is Raman active. In the bending vibration, the bond length does not change, and in the asymmetric stretching vibration, the effects of both C=O bonds cancel each other out. For CO_2, one would, therefore, expect a Raman spectrum consisting of just one line. Accidentally, the frequency of this Raman band falls close to twice the frequency of the bending vibration (667 cm^{-1}). Thus, Fermi resonance occurs, leading to two Raman peaks of similar intensity at 1388 and 1285 cm^{-1}.

4.2.5   Spurious Bands in Infrared Spectra

Traces of water in carbon tetrachloride and chloroform may give rise to bands near 3700 and 3600 cm^{-1}. In addition, a weak broad band is observed around 1650 cm^{-1}. Water vapour gives many sharp bands between 2000 and 1280 cm^{-1}. If present in relatively high concentrations, these bands may be the source of artifacts, which are particularly annoying when scanning through a steep flank of a strong peak. Sometimes, puzzling shoulders on carbonyl absorption bands can be explained by this effect.

Dissolved carbon dioxide causes an absorption band at 2325 cm^{-1}. In solutions containing amines, the dissolved carbon dioxide, together with the ubiquitous water, may form carbonates. As a consequence, the spectrum exhibits unexpected bands due to the protonated amine function. In the vapour phase, the absorption of carbon dioxide may be the cause of increased base line noise at 2360 and 2335 cm^{-1}, which is accompanied by another band at 667 cm^{-1}.

Commercial polymeric materials often contain phthalates as plasticizers, which sometimes find their way into "pure" analytical samples. They give rise to a band at 1725 cm^{-1}. Various chemical operations on the sample may transform the phthalates into phthalic anhydride, which gives an absorption at 1755 cm^{-1}.

Another rather common impurity are the various silicones. These usually result in an absorption at 1265 cm^{-1}, together with a broad band in the 1100–1000 cm^{-1} region. If carbon tetrachloride evaporates from a leaky cell, a band at 793 cm^{-1} is observed. However, for liquid carbon tetrachloride, the corresponding absorption appears at 788 cm^{-1}. This band is observed if the outside of the cell is contaminated with carbon tetrachloride.

4.3 NMR Spectroscopy

4.3.1 Presentation of Spectra

4.3.1.1 ^{1}H and ^{13}C NMR Spectra

The original ^{1}H and ^{13}C NMR spectra are reproduced throughout. If not otherwise stated, all one- and two-dimensional spectra have been recorded in 5-mm-sample tubes, which were not rotated, at room temperature using a Bruker DRX-500 spectrometer (11.7 Tesla; 500 MHz for ^{1}H NMR and 125 MHz for ^{13}C NMR). Chemical shifts are given on the δ scale (δ$_{TMS}$ = 0.00). For the ^{1}H NMR spectra, tetramethylsilane was applied as internal reference, whereas for ^{13}C NMR spectra, the line of the solvent was used as an indirect reference using the following shift values: CDCl$_3$: δ 77.0, DMSO-d$_6$: δ 39.5, and CD$_3$OD: δ 49.0. In the one-dimensional measurements, 64 K data points were used for the FID. In many cases, Gaussian-type resolution enhancement was applied for the ^{1}H NMR spectra. For the the two-dimensional measurements, the data matrix had the dimension 2 K × 256 K. For processing, linear prediction to 512 K and zero filling up to 1 K data points were utilized in the F1 dimension. On the top and the left of the two-dimensional spectra, the corresponding one-dimensional ^{1}H or ^{13}C NMR spectra are shown.

The following kinds of spectra were recorded: one-dimensional ^{1}H and ^{13}C NMR, DEPT135, and NOESY, as well as two-dimensional ^{1}H,^{1}H COSY, NOESY/EXSY, one-bond correlated ^{13}C,^{1}H COSY (HSQC or HMQC), and ^{13}C,^{1}H COSY optimized for a coupling constant, $J_{C,H}$ = 7 Hz (HMBC). In the case of one-dimensional NOESY and all COSY experiments, the more effective gradient-selected method was applied to suppress unwanted signal distortions and select the required coherence path.

The DEPT (Distortionless Enhancement by Polarization Transfer) experiment uses polarization transfer from protons to carbon nuclei resulting in considerable sensitivity gain and may be applied as a powerful means for distinguishing CH$_3$, CH$_2$, and CH groups. The DEPT135 version applied here, gives edited spectra where the signals for CH$_3$ and CH appear upside and for CH$_2$ downside, those of the quaternary carbons being absent. Ideally, one pulse delay, τ, should exactly match $1/(2J)$, J being the ^{13}C,^{1}H coupling constant over one bond. However, the method is not very sensitive to deviations from this ideal value as long as they are not too large (such as in alkynes). For the present edition, all DEPT135 spectra were recorded with τ = 3.45 ms, corresponding to $^1J_{C,H}$ = 145 Hz.

Irradiation of a proton group causes a change in the signal intensities of other protons having a dipolar coupling with them (so-called Nuclear Overhauser Effect, NOE). The dipolar coupling is related to the inverse sixth power of the distance between the nuclei. Thus, the NOE provides information about the spatial proximity of nuclei in a molecule. The one-dimensional NOESY (Nuclear Overhauser Enhancement and exchange Spectroscopy) experiment yields a clear high resolution spectrum, which only contains signals of those protons that are in sterical neighborhood (≤ ca. 5 Å) to the irradiated proton. The two-dimensional, phase-sensitive NOESY experiment gives cross peaks for protons located in sterical proximity with an opposite phase to those of the diagonal signals. In this experiment, the digital resolution is limited, especially in the F1 dimension. Both methods are interpreted in the same qualitative way: An interaction is expected for distances ≤ ca. 5 Å between the nuclei.

The 2D NOESY pulse sequence, additionally, yields cross peaks with the same phase as the diagonal signals among all exchanging species. This EXSY (Exchange SpectroscopY) spectrum indicates chemical exchange, even if it is so slow that no line broadening can be detected in the one-dimensional ^{1}H NMR spectrum. The EXSY cross signals can be very strong, displaying an intensity never reached by NOE signals.

The COSY (COrrelation SpectroscopY) pulse sequence generates two-dimensional spectra in which cross peaks appear among scalar coupled nuclei. The ^{1}H,^{1}H COSY measurement detects coupled pairs of protons, allowing the determination of connectivities over two or three bonds.

To obtain higher sensitivity, the so-called inverse detection, i.e., the ^{1}H-detected modes, Heteronuclear Single Quantum Correlation (HSQC) and Heteronuclear Multiple Quantum Correlation (HMQC), were applied for the measurement of heteronuclear ^{13}C,^{1}H COSY spectra. The cross peaks in these spectra show C,H pairs having a coupling over one bond ($^1J_{C,H}$). The HSQC method yields better signal to noise ratios but is very sensitive to small errors in the setting of pulse lengths, whereas the HMQC is more robust. An interesting and useful mode is the ^{13}C-coupled HSQC spectrum yielding doublets in the ^{1}H (F2) dimension so that the $^1J_{C,H}$ coupling constants can be obtained directly from this measurement. The HSQC and HMQC are largely limited to one-bond couplings and do not provide information on quaternary carbons. The Heteronuclear Multiple Bond Correlation (HMBC) measurement, on the other hand, enables one to make assignments in cases where ^{1}H and ^{13}C nuclei are coupled through two or more bonds with smaller coupling constants. In such spectra, one-bond couplings are suppressed, however, this is strictly valid only if the timing fits their value. The parameters of all HMBC experiments in this book are set to obtain optimal cross peaks with $J_{C,H} = 7 \pm (3-4)$ Hz. The standard setting of $1/(2J) = 3.45$ ms corresponds to $^1J_{C,H} = 145$ Hz. If the actual $^1J_{C,H}$ strongly differs from this value, a doublet appears in the two-dimensional spectrum, which allows to also determine the $^1J_{C,H}$ value. Occasionally, the coupling constants over four bonds are sufficiently large (> ca. 2 Hz) to be observed in the HMBC spectra.

4.3.1.2 Nuclear Overhauser Effect

The dipolar couplings and, thus, the intensity of NOE signals strongly depend on the rotational mobilities. The NOE is positive for small molecules, which rapidly rotate in solution; its magnitude decreases with increasing molecular weight, passes through zero, and then becomes negative (the limiting value being ca. −100%) if the mobility is further reduced. Molecules with M_r ca. 1 000 u are likely to fall within the intermediate region where no NOE can be detected. In such cases, the related Rotating Frame Nuclear Overhauser Enhancement SpectroscopY (ROESY) may be applied since with this method, the dependence of the NOE effect on the molecular rotational mobility is negligible.

4.3.2   Rules for the Interpretation of Coupling Patterns

4.3.2.1   Definitions

It is useful to define isochronicity (chemical shift equivalence) and magnetic equivalence (chemical shift and coupling constant equivalence) as given below. The term "chemical equivalence" which is often used erroneously in this context should be avoided because it is not precisely defined and, in several respects, not directly relevant to NMR spectroscopy.

Isochronous nuclei

Nuclei are isochronous if there is no measurable difference in their chemical shifts under the given experimental conditions. Isochronicity may be a consequence of molecular symmetry, fast intra- or intermolecular exchange, or it can be purely accidental. Except for homotopic nuclei (see Section 4.3.4.2), isochronous nuclei may become nonisochronous when the experimental conditions are altered.

Magnetic equivalence

Nuclei are magnetically equivalent if they are isochronous and if they exhibit identical coupling constants with every individual member of the set of all other nuclei. The members of an isolated set of isochronous nuclei, which are not coupled to any other nuclei, are also magnetically equivalent. When the equivalence of chemical shifts and coupling constants is not a consequence of rotational symmetry, magnetically equivalent nuclei may become nonequivalent if the experimental conditions are changed.

Nomenclature of spin systems

A spin system extends over a set of nuclei which are connected by an unbroken coupling path. If there are two sets of nuclei in a molecule where no member of one group couples with any member of the other, the two sets can be treated as independent spin systems. Two different nomenclature systems are widely used, but only the one employed in this volume is presented here.

A set of magnetically equivalent nuclei is denoted by a capital letter with an index giving the number of magnetically equivalent nuclei within the group. Isochronous nuclei that are not magnetically equivalent are denoted by the same capital letter, but are distinguished by primes. Nonisochronous nuclei are denoted by letters adjoining in the alphabet if they are strongly coupled so that higher-order spectra result (chemical shift difference small relative to coupling constant, i.e., Δv [Hz] < ca. 10 J [Hz]). If the spins are weakly coupled (the chemical shift differences are large, i.e., Δv [Hz] > ca. 10 J [Hz]), the nuclei are symbolized with letters far apart in the alphabet. Note that higher-order spectra can often be simplified by suitably changing the experimental conditions, in particular by recording the spectrum at higher magnetic field strength. However, the presence of isochronous but magnetically nonequivalent nuclei with nonzero coupling will always lead to higher-order spectra (their chemical shift difference is zero by definition).

4.3.2.2 General Rules

The following rules apply both to first- and higher-order spectra.
1. Coupling between magnetically equivalent nuclei does not affect the spectrum and is, therefore, not defined by the spectrum.
2. The coupling constants do not depend on the strength of the applied magnetic field. Since, on the other hand, the chemical shift differences measured in Hz depend on it, spin systems leading to higher-order spectra at a given magnetic field may change to first-order spectra at a higher magnetic field strength.
3. Each spin-spin interaction is mutual: If a nucleus A is coupled with another nucleus B (or X), then B (or X) is coupled with A to exactly the same extent: $J_{A,B} = J_{B,A}$.

4.3.2.3 First-Order Spectra

First-order spectra are observed if the chemical shift differences between all magnetically nonequivalent nuclei within a spin system are large relative to the corresponding coupling constants. If there is no coupling between two sets of nuclei, they constitute two independent spin systems. In practice, as a rule of thumb, first-order spectra may be expected if the following condition applies:

$$\Delta\nu_{ij} > \text{ca. } 10\, J_{ij} \tag{4.8}$$

for all pairs of magnetically nonequivalent nuclei in a spin system. If a spin system exhibits a higher-order part, the higher-order effects will, in general, complicate the signals of the remaining nuclei as well.

For isochronous ($\Delta\nu_{ij} = 0$) but magnetically nonequivalent nuclei, higher-order spectra are to be expected regardless of the value of the coupling constants between the isochronous nuclei (e.g., $AA'XX'$ spin systems generally consist of 20 lines, whereas first-order rules would predict only 8 lines). Such a spin system can never be simplified by applying higher magnetic field strengths.
1. The multiplicity of the signal of A in an A_mX_n spin system is determined by the number n and spin quantum number I of the nuclei X by the relation $(2nI + 1)$. For protons, carbon-13, and other nuclei with $I = 1/2$, this amounts to $(n + 1)$. If there are more than two interacting groups such as in an $A_nM_mX_p$ system of nuclei with $I = 1/2$, the multiplicity of the signal A will, in general, be given by $(m + 1)(p + 1)$. For special ratios of the coupling constants, some lines may coincide so that the number of observed lines may be reduced. If, for example, $J_{A,M} = J_{A,X}$ in the above case, then $(m + p + 1)$ lines will be observed.
2. The intensities and positions of the lines are symmetric about the center of the multiplet, which corresponds to the chemical shift.
3. If a multiplet is produced by coupling with a single group of equivalent nuclei, a multiplet of equidistant lines will be produced whose spacing is equal to the coupling constant J. If this group consists of n equivalent nuclei with $I = 1/2$, the relative intensities of the lines within the multiplet are given by the coefficients of the

binomial expansion, i.e., 1:1 ($n = 1$), 1:2:1 ($n = 2$), 1:3:3:1 ($n = 3$), 1:4:6:4:1 ($n = 4$), 1:5:10:10:5:1 ($n = 5$), 1:6:15:20:15:6:1 ($n = 6$), etc. The corresponding relative intensities from coupling with n equivalent nuclei having $I = 1$ are 1:1:1 ($n = 1$), 1:2:3:2:1 ($n = 2$), 1:3:6:7:6:3:1 ($n = 3$), etc.

4. It is often useful to rationalize the splitting and intensities of the signals by drawing the splitting diagrams. It is of no importance in which order the individual couplings are considered. Thus, the two representations shown in Fig. 4.1 for the X part of a hypothetical A_2MX system are equivalent.

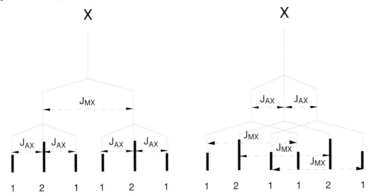

Fig. 4.1: Two equivalent ways of rationalizing the splitting for the X part of a hypothetical A_2MX system

5. Since pure first-order spectra would exist only if the ratio $\Delta v/J$ were infinitely large, the first-order analysis for homonuclear spin systems leads to minor errors. The splitting and positions of the lines are, generally, not affected if $\Delta v/J >$ ca. 10 but their intensities are different from those predicted by first-order rules. If for two lines the same intensities are predicted by first-order rules, the one being closer to the corresponding coupling partner becomes more and the other one less intense than predicted. This effect is very useful in practice because the connecting lines between peak tops are pointing in the direction of the coupling partners (see Fig. 4.2).

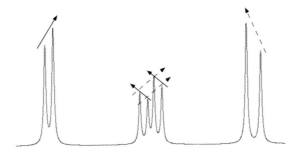

Fig. 4.2: Intensity differences between signals of a multiplet indicate the relative chemical shift of the corresponding coupling partner

4.3.2.4 Higher-Order Spectra

If nuclei in a spin system are strongly coupled, i.e., if $\Delta v_{ij} < 10\ J_{ij}$, higher-order spectra will be observed, which generally consist of more lines than predicted by the rules of first-order analysis. The set of parameters (chemical shifts and coupling constants) are, in general, not easily extracted from such spectra. If such higher-order spectra are encountered, the chemist has the following possibilities of interpreting the data:

1. Full analysis, i.e., determination of all available parameters by computer simulation of the spectra.
2. Recording of the spectrum under changed experimental conditions, such as spin decoupling, other solvents, higher magnetic field strength, isotopic substitution, or application of shift reagents, in order to obtain first-order spectra.

In many cases, qualitative features, i.e., information about the number of protons and the approximate chemical shift values, are sufficient for solving the problems, and no further analysis is necessary. Sometimes, misleadingly simple-looking spectra are observed ("deceptive simplicity") and interpretation according to first-order rules gives erroneous results. For example, *ABX* spin systems often lead to such errors. The *X* part consists of six lines in the general case but often only four lines are discernible. One may easily be tempted to interpret this part as a first-order spectrum (*X* part of an *AMX* system) and assign the *AX* and *BX* coupling constants to the splittings. In general, however, this leads to false results since the line splittings are not always equal to the coupling constants in an *ABX* spectrum. By the same token, even if only a triplet is observed for the *X* part, the coupling constants $J_{A,X}$ and $J_{B,X}$ may be different. On the other hand, if the *X* part consists of four lines, it positively indicates that $J_{A,X} \neq J_{B,X}$. In general, δ_X and $|J_{A,X}+J_{B,X}|$ (distance between the outer lines in the four-line pattern) are directly accessible from the *X* part of the spectrum, whereas $J_{A,B}$ (the spacing found four times in the *AB* portion) is the only parameter directly obtainable from the *AB* part of *ABX* spectra.

Very useful qualitative information may be gathered from the presence or absence of symmetry in the spectra of four-spin systems. The *AA'BB'*, *AA'XX'*, and A_2B_2 (in practice rarely occurring) systems always give rise to a symmetrical pattern. The first two systems occur in benzene if it is *p*-disubstituted with different substituents or *o*-disubstituted with equal substituents. Although both types give rise to symmetrical patterns, the general shape of the spectra is not the same so that a differentiation is, generally, possible. The spectra of *p*-disubstituted benzenes have some resemblance to *AB*-type spectra, while this is not true for *AA'BB'* systems obtained from *o*-disubstituted benzenes with two identical substituents.

4.3.2.5 Computer Simulation

A set of n chemical shifts and $n(n-1)/2$ coupling constants unambiguously defines a spectrum. Several computer programs are available for calculating spectra on the basis of exact solutions of quantum mechanical equations. Quite often, approximate interpretation according to first-order rules is possible but such an interpretation should be verified by computer simulation.

4.3.3   Signal Intensities in ^{13}C NMR Spectra

If a system is in thermal equilibrium, the signal intensities in NMR spectra are proportional to the number of absorbing nuclei. In general, ^{13}C NMR spectra are recorded under experimental conditions which cause a nonequilibrium population of spin states. In this section, the most important factors influencing the line intensities in routine ^{13}C NMR spectra are discussed.

4.3.3.1   Saturation

Today, ^{13}C NMR spectra are exclusively recorded using the pulse Fourier-transform technique where a short and intense pulse is applied to excite all ^{13}C nuclei present. The response of the system, the interferogram, containing all frequencies of the spectrum is then recorded. If a 90^0 pulse was applied, which leads to the maximum intensity following a single pulse, a nucleus reaches 99.98% of the equilibrium magnetization after a time of 5 T_1 s, where T_1 is the spin-lattice relaxation time of the nucleus. In practice, a great number of interferograms are recorded whereby the signal to noise ratio is improved by a factor of $\sqrt{n}$ if n interferograms are accumulated. The pulse rate is, generally, kept higher than necessary for complete relaxation of all nuclei present. The interferograms are thus recorded under conditions of partial saturation. Nuclei with longer relaxation times are more extensively saturated (their spin states are further away from the equilibrium) and, thus, give rise to reduced signal intensities, while nuclei with the same relaxation time always yield the same signal intensity regardless of the extent of saturation.

With the exception of small molecules and of some quaternary carbon atoms in organic compounds, practically all ^{13}C nuclei relax by the way of dipole-dipole interactions with protons. The most important factors influencing the dipole-dipole relaxation time are the number of protons attached to a carbon atom and the rotational mobility of the C–H vectors. In most cases, the following sequence of relaxation times is observed:

$$CH_2 < CH < CH_3 << C \tag{4.9}$$

The relaxation time of methylene carbon atoms is shorter than that of methine carbon atoms because of the higher number of directly attached protons. Methyl carbon atoms exhibit, in general, longer relaxation times than methylene and methine carbon atoms as a consequence of their higher rotational mobility. Since the magnetic dipole-dipole interaction is dependent on the sixth power of the internuclear distance, quaternary carbon atoms show much longer relaxation times than protonated carbon atoms. The sequence of expected signal intensities caused by partial saturation is thus:

$$CH_2 \geq CH \geq CH_3 \geq C \tag{4.10}$$

Equal intensities are only expected if no saturation occurs and no other factors influence the line intensities. The relaxation times can be reduced by the application of paramagnetic relaxation reagents such as chromium(III) acetylacetonate.

4.3.3.2 Nuclear Overhauser Effect

The equilibrium population and, therefore, the line intensity of a nucleus can be influenced by irradiation at the Larmor frequency of another nucleus. Such effects, which occur in double resonance experiments if the nuclei are coupled by dipolar interactions, are termed nuclear Overhauser effect (NOE).

Since, in most cases, ^{13}C nuclei relax through dipolar interactions with protons, proton decoupling influences the carbon-13 line intensities. The intensities I thereby increase by a factor, η, which depends on the relaxation mechanism of the respective ^{13}C nucleus:

$$I = 1 + \eta = 1 + 2\frac{T_1}{T_{1CH}} \tag{4.11}$$

where T_{1CH} is the $^{13}C-^{1}H$ dipolar relaxation time and T_1 the overall spin-lattice relaxation time:

$$\frac{1}{T_1} = \frac{1}{\sum_i T_{1i}} \tag{4.12}$$

with T_{1i} as the relaxation time due to the i-th mechanism.

If the $^{13}C,^{1}H$-dipolar interaction is the only effective relaxation mechanism, T_1 becomes equal to T_{1CH} and the line intensity is three times higher than it would be without NOE. If, on the other hand, other mechanisms dominate, T_1 will be much smaller than T_{1CH} and no NOE occurs. Relaxation reagents, therefore, reduce the NOE.

The shorter the $^{13}C,^{1}H$-dipolar relaxation time, the smaller is the probability of other mechanisms being significant. The order of intensity enhancements by NOE, thus, exactly corresponds to the sequence (4.10) given above for the expected line intensities due to partial saturation.

If full NOE occurs for all nuclei, the relative line intensities are not influenced. It is to be noted that equation (4.11) only applies if the molecular rotational mobilities are high relative to the reciprocal of the resonance frequencies (extreme narrowing condition). This is practically always the case for small- or medium-sized organic molecules in common solvents at room temperature if low or medium magnetic field strength is applied. At high magnetic field strength of 10 Tesla or more (i.e., > ca. 100 MHz for ^{13}C), the extreme narrowing condition may not be met even for medium-sized organic molecules. This leads to a reduced NOE even where the $^{13}C,^{1}H$-dipolar interaction is the predominant relaxation mechanism. Particular care must, therefore, be taken when interpreting line intensities in such cases.

Elimination of NOE can be easily achieved by applying the decoupling field in a pulsed mode. Since the multiplets collapse almost instantaneously when the decoupling field is turned on, whereas the NOE builds up with the time constant of the order of T_1, the NOE may be suppressed by gating the decoupler so that it is on only during each data acquisition period but off during each pulse delay.

4.3.3.3 Intensities of Solvent Signals

Generally, deuterated solvents are applied in ^{13}C NMR spectroscopy since a deuterium reference signal is used for the stabilization (lock) of the magnetic field. The relaxation times of the ^{13}C nuclei in deuterated solvents are very long for several different reasons. First of all, the usual solvents are small and highly mobile molecules, leading to less effective dipolar interaction. Secondly, the dipolar relaxation with deuterium is much less effective than with a H atom. The pulse rates usually applied thus cause extensive saturation of the solvent signals. Furthermore, in nonprotonated molecules, the NOE is absent. Finally, coupling with deuterium leads to line splittings and, thereby, to a further reduction of the intensity of individual lines.

4.3.4 Influence of Molecular Symmetry and Conformational Equilibria on NMR Spectra

4.3.4.1 Introduction

The influence of symmetry and fast conformational equilibria on the NMR spectra very often causes difficulties in practical analysis which have led to numerous erroneous statements in the literature and even in textbooks. In this section, we first present a consistent set of useful definitions and then give rules to predict the type of the expected spin system on the basis of symmetry considerations. Specific examples are then given to amplify the use of these rules. The prediction of the type of the expected spin system is often sufficient for spectra interpretation and it is also the first step if computer simulations are performed to evaluate the exact parameters.

4.3.4.2 Definitions

Relationships between pairs of nuclei of the same kind in a given structure

To predict isochronicity, the symmetry relation between two nuclei of the same kind is of relevance. Symmetry relations always refer to one specified structure, i.e., to fixed positions of all nuclei. Different conformations of a molecule define different structures.

Nuclei are *constitutionally equivalent* if they have the same connectivity (bondedness).

Nuclei are *diastereotopic* if they are constitutionally equivalent but not equivalent by symmetry.

Nuclei are *enantiotopic* or equivalent by reflection if the structure has an improper axis by which the nuclei are interchanged but no corresponding proper axis. Note that a symmetry plane and an inversion center are special cases of improper axes.

Nuclei are *homotopic* or equivalent by rotation if the structure exhibits a proper axis by which the nuclei are interchanged.

Relationships between one nucleus and a pair of nuclei of the same kind in a given structure

To predict the equivalence of two coupling constants, the symmetry of the coupling paths is relevant. Although the coupling is mainly a through-bond interaction, for the present purpose, the coupling path can be considered as a connecting line (edge) between the respective nuclei. To consider the relation between two nuclei and a selected coupling partner, we consider the relation of the corresponding connecting lines. Again, the symmetry relation can be nonsymmetric, symmetric by reflection, or symmetric by rotation. In the last two cases, the coupling partner must be located on the symmetry element (i.e., on the plane for improper symmetry operations or on the proper axis for proper rotations).

4.3.4.3 Influence of Symmetry Properties on the NMR Spectra

Homotopic nuclei are always isochronous and homotopic coupling paths always lead to equal coupling constants. In a nonchiral environment (including solvent, complexing agent, shift reagents, etc.), enantiotopic relationships induce equivalence of the chemical shifts or coupling constants. In a chiral environment, enantiotopic relationship does not imply magnetic equivalence. A chiral environment has, however, only a small effect on coupling constants. Thus, loss of magnetic equivalence due to enantiotopic coupling paths in a chiral environment has not yet been observed experimentally. Of course, equivalence may occur by chance in any other case. For example, diastereotopic nuclei are often isochronous if the chiral part of the molecule is far away, e.g., if a chirality center is at a distance of at least three to four bonds.

4.3.4.4 Influence of Fast Equilibria

If chemical equilibria are fast, i.e., the mean lifetime of the species is small as compared to the relevant reciprocal chemical shift differences (measured in [Hz]), a single line with an average chemical shift is observed:

$$\delta = \sum_i \frac{c_i \delta_i}{c_{tot}} \tag{4.13}$$

where c_{tot} is the total concentration summed over all environments, c_i the relative concentration in the environment i, and δ_i is the respective chemical shift. The c_i are the concentrations of the nuclei, not of the species. Thus, in a 1:1 molar mixture of water and acetic acid, c_{water}/c_{COOH} is 2:1. An analogous relationship holds for the coupling constants in the case of a fast intramolecular exchange (however, no couplings between the respective nuclei are observed if the fast exchange is intermolecular). By applying this equation, the parameters can be calculated, provided their values in the various environments as well as the concentrations are known. The application of this procedure for predicting the chemical shifts of a CH_3 and a CH_2 group in the hypothetical chiral compounds, CH_3CXYZ (Fig. 4.3) and RCH_2CXYZ (Fig. 4.4), is shown in the following paragraphs.

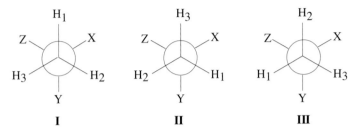

Fig. 4.3: Conformations considered for predicting the chemical shifts of the methyl protons in a hypothetical compound CH_3CXYZ (X, Y, Z are different substituents)

The chemical shifts of the three protons if fast exchange according to eq 4.13 occurs, are given in eqs 4.14–4.16 (c_i denotes the individual concentrations, c_{tot} is their sum):

$$\delta_{H_1} = \frac{c_I}{c_{tot}}\delta_{H_1(I)} + \frac{c_{II}}{c_{tot}}\delta_{H_1(II)} + \frac{c_{III}}{c_{tot}}\delta_{H_1(III)} \tag{4.14}$$

$$\delta_{H_2} = \frac{c_I}{c_{tot}}\delta_{H_2(I)} + \frac{c_{II}}{c_{tot}}\delta_{H_2(II)} + \frac{c_{III}}{c_{tot}}\delta_{H_2(III)} \tag{4.15}$$

$$\delta_{H_3} = \frac{c_I}{c_{tot}}\delta_{H_3(I)} + \frac{c_{II}}{c_{tot}}\delta_{H_3(II)} + \frac{c_{III}}{c_{tot}}\delta_{H_3(III)} \tag{4.16}$$

In reality, the conformers **I–III** are indistinguishable because they can be derived from each other by rotating the methyl group around a bond, which represents a threefold symmetry axis for this group. As a consequence, eqs 4.17–4.20 are valid:

$$c_I = c_{II} = c_{III} = \frac{1}{3}c_{tot} \tag{4.17}$$

$$\delta_{H_1(I)} = \delta_{H_3(II)} = \delta_{H_2(III)} \tag{4.18}$$

$$\delta_{H_2(I)} = \delta_{H_1(II)} = \delta_{H_3(III)} \tag{4.19}$$

$$\delta_{H_3(I)} = \delta_{H_2(II)} = \delta_{H_1(III)} \tag{4.20}$$

By combining eqs 4.17–4.20 with eqs 4.14–4.16, it becomes evident that:

$$\delta_{H_1} = \delta_{H_2} = \delta_{H_3} \tag{4.21}$$

Thus, the three protons of a methyl group in a chiral environment are shown to be isochronous.

For methylene protons, the situation is different:

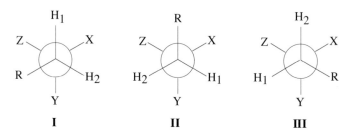

Fig. 4.4: Conformations considered for predicting the chemical shifts of the methylene protons in a hypothetical compound RCH_2CXYZ (R, X, Y, Z are different substituents)

The chemical shifts of the two protons in fast exchange according to eq 4.13 are given in eqs 4.22 and 4.23 (c_i stands for the individual concentrations, c_{tot} is their sum):

$$\delta_{H_1} = \frac{c_I}{c_{tot}} \delta_{H_1(I)} + \frac{c_{II}}{c_{tot}} \delta_{H_1(II)} + \frac{c_{III}}{c_{tot}} \delta_{H_1(III)} \tag{4.22}$$

$$\delta_{H_2} = \frac{c_I}{c_{tot}} \delta_{H_2(I)} + \frac{c_{II}}{c_{tot}} \delta_{H_2(II)} + \frac{c_{III}}{c_{tot}} \delta_{H_2(III)} \tag{4.23}$$

In the general case, the concentrations of the individual conformers are not equal because the axis of rotation is not a local symmetry axis for the CH_2R group. Thus, the substituent R has different *gauche* substituents in each case:

$$c_I \neq c_{II} \neq c_{III} \tag{4.24}$$

Furthermore, there is no reason for an equivalence of chemical shifts of protons in virtually the same position (e.g., H_1 in **I** and H_2 in **III**). This is easily seen if one assumes, for example, a bulky substituent Y, in which case the idealized conformations as shown in Fig. 4.4 are not exactly valid: To reduce vicinal steric interactions, the CH_2R group will be pushed clockwise in **I** and counterclockwise in **III** by the interaction of the bulky substituent Y with R so that the exact positions of H_1 in **I** and H_2 in **III** are different:

$$\delta_{H_1(I)} \neq \delta_{H_2(III)} \tag{4.25}$$

$$\delta_{H_2(I)} \neq \delta_{H_1(II)} \tag{4.26}$$

$$\delta_{H_1(III)} \neq \delta_{H_2(II)} \tag{4.27}$$

Thus, it is obvious that the chemical shifts of diastereotopic methylene protons cannot become equal owing to fast rotation.

A treatment as shown above is only rarely needed in practice. For fast conformational equilibria, the type of spectra can be predicted by applying the rules given below. They are, however, not valid for other kinds of fast exchange reactions.

For predicting the type of spectra in molecules with fast conformational equilibria, first, atom groups are identified that relative to the rest of a molecule are rotating rapidly around a bond, which is also a symmetry axis for this group. All nuclei permutated by this rotation become isochronous and, with respect to all couplings with nuclei in the rest of the molecule, equivalent. Thus, the protons of a methyl group are always magnetically equivalent. The deeper reason for the validity of this rule is, as shown above, that due to the local symmetry, the populations of the various conformers as well as the chemical shifts of the different nuclei in the same position will be the same.

For the rest of the molecule with a fast conformational equilibrium, the rules described above for a rigid system may then be applied to an arbitrarily chosen conformation of the highest possible symmetry, regardless of whether this conformation is energetically reasonable. For example, the planar conformation of cyclohexane (!) is appropriate. The reason for this rule is that any other conformation will have a symmetrically equivalent counterpart showing the same population so that the corresponding parameters will average out as if the molecule were in the single symmetrical conformation.

4.3.4.5 Examples

Molecule	Type of spectrum in nonchiral environment	Type of spectrum in chiral environment	Remarks
	A_2X_2	$AA'XX'$	The ^1H and ^{19}F nuclei are pairwise homotopic but the relevant coupling paths are only enantiotopic
	A_2X_2	$AA'XX'$	Topologically the same as the previous example
	$AA'BB'$ or $AA'XX'$	$AA'BB'$ or $AA'XX'$	The protons are pairwise homotopic but the relevant coupling paths (i.e., one *ortho* and one *para* coupling) are not related by symmetry

Molecule	Type of spectrum in nonchiral environment	Type of spectrum in chiral environment	Remarks
	AA'XX'	*AA'XX'*	Topologically the same as the previous example
	AA'BB' or *AA'XX'*	*AA'BB'* or *AA'XX'*	Though the molecule is chiral, fast rotation around the $Ph–CR_1R_2R_3$ axis leads to pairwise isochronicity of the aromatic protons
	AA'BB' or *AA'XX'*	*ABCD* or *ABXY*	The geminal protons are enantiotopic and there is no symmetry relationship between the vicinal couplings. In practice, the spin system often mimics an A_2X_2 one
	A_2BC	*ABCD*	The methylene protons are enantiotopic
	A_2BC or A_2XY	*ABCD* or *ABXY*	The methine protons are enantiotopic, the methylene protons diastereotopic
	AA'BB' or *AA'XX'*	*AA'BB'* or *AA'XX'*	Both the methylene and the two methine protons are homotopic but there is no symmetry relation between the relevant coupling paths

5 Assemble 2.1 Tutorial

5.1 Brief Description

Assemble 2.1 is a structure- or, more precisely, a constitution-generating program. It always requires a molecular formula. A variety of structural features can be specified. It is important to note that Assemble only accepts structural, not spectral, information. The program does not interpret any spectral features. It is a purely mathematical tool. All interpretation tasks and, consequently, all responsibility for their correctness is left to the user. However, Assemble allows the user to enter the structural information in the form as it is derived from spectra.

5.2 Introductory Example

When you start up the program, the main input window of Assemble 2.1 appears.

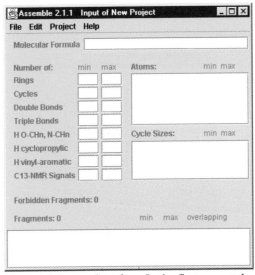

This is the central part of the user interface. In the first row, at the top of the window, you enter the molecular formula. This is peremptory. Assemble will not run without this item.

To start with a simple example, enter the molecular formula of $C_7H_{14}O$ by clicking into the text field and typing the formula. If no further input is given, Assemble will generate all constitutional isomers corresponding to that molecular formula. Assemble knows a number of structural features that lead to molecules too strained to be stable. For each of the strained features, you can choose whether you want to have it detected. From the "Edit" pull-down menu choose "Forbidden Fragments".

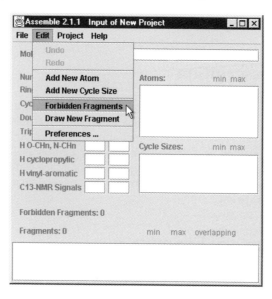

After releasing the mouse button, a window pops up showing the various strained features.

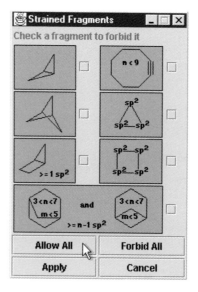

By clicking the field "Allow All" at the bottom of the window, the strained features detector will be switched off. When you click the "Apply" field, the window vanishes. To start the structure generator, use the "Project" pull-down menu of the main input window. Select "Calculate with BC ...".

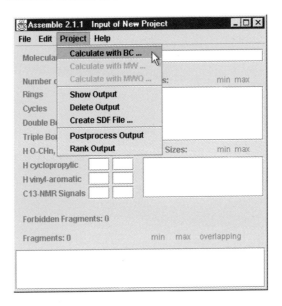

The input to Assemble is stored on disk for later reference. Have a new project created by clicking the corresponding button. Choose or type a project name. You can reload the input later when selecting the project by its name.

Assemble makes certain obvious assumptions about the valences of the elements. Hydrogen and carbon atoms have fixed valences of 1 and 4, respectively. The valences of hetero atoms can be altered. After entering a formula with an oxygen atom, you can set the valence to a value different from the default value of 2. Leave it as it is and click into the "OK" bar. After a few seconds, a new window will pop up displaying the first nine structures generated out of a total of 596 isomers, as you can see from the title bar of the window.

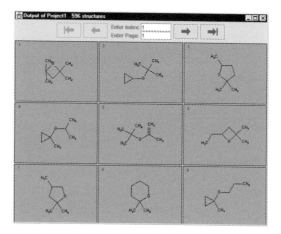

Click into the "Next Page" field to look at further structures.

To generate all isomers corresponding to a molecular formula is of little use. If you are in a process of structure elucidation, you certainly have additional information from various sources to restrict the number of candidate structures generated. By looking at the main input window, you quickly get an idea of how to enter some of this information. For several features, a minimum and maximum value can be specified.

Assume that you know there is a double bond in your molecule. You may have gathered this from the IR spectrum or from the appropriate chemical shifts in the ^{13}C NMR spectrum. You can enter this information by clicking into the text field "min" of the "Double Bonds" row and type 1.

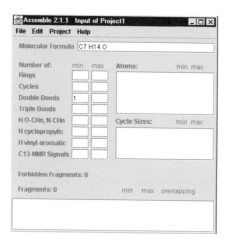

If you want to specify a maximum value, you can do so but there is no need for it in this example since the molecular formula corresponds to a single double bond equivalent, which can be a double bond or a ring. Leaving the "max" field blank does not impose an upper limit. When you now restart the generator as shown before, you are first asked where to save the input. Choose a new project name or have the latest version of the current project overwritten.

After starting the generator, the structures appear in a new window. All of the 294 isomers generated have a double bond, as requested.

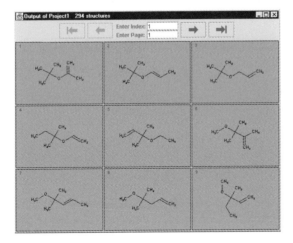

The multiplicity of the signals in the off-resonance ^{13}C NMR spectrum yields information on the number of hydrogens immediately bonded to a carbon atom. In addition, the hybridization of the atom may be derived from the chemical shift of the signal. Similarly, you may know how many OH groups there are in the molecule by interpreting the integral of their signal in the ^{1}H NMR spectrum. This kind of information is most easily entered as so-called atom constraints in the text field "Atoms:". When you click into this field with the right mouse button, a pop-up menu appears. Select the "Add New Atom" line.

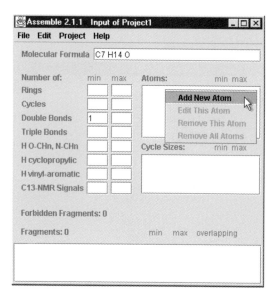

A new input window appears where you can specify the element, the number of hydrogen atoms directly bonded to it, its hybridization, and the minimum and maximum number of such atoms in the molecule.

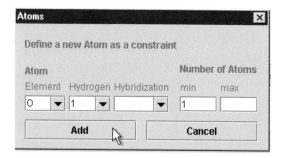

Clicking the "Add" button will add a line in the "Atoms:" field of the main input window. As an example, request the presence of an OH group. You enter the element symbol O in the first field, 1 in the second for the number of hydrogens, leave blank the hybridization field (or clear it if there already is an entry), then enter the minimum value of 1, and omit the maximum. After clicking the "Add" button, you will see a new line in the "Atoms:" text field.

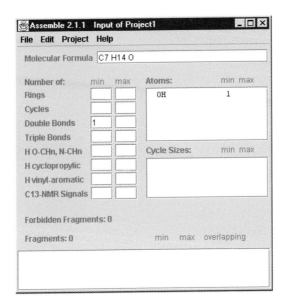

After starting the generator, a total of 149 structures are created, each one with an OH group and a double bond.

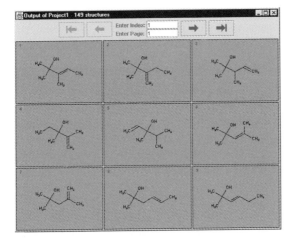

If you know that the molecule under study has a fragment comprising more than one non-hydrogen atom, you cannot enter them as atom constraints. For example, assume the presence of an ethyl group. The large field in the bottom part of the main input window is designed for larger fragments. Click into this field with the right mouse button and a pop-up menu will appear. Select the line "Draw New Fragment".

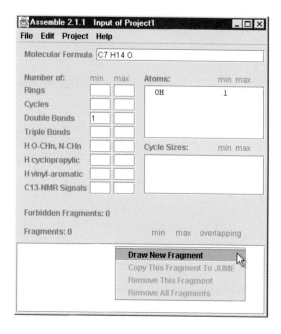

Drawing substructures with the program JUME is straightforward. You best learn it by practicing. Have a look at the help text by clicking at the question mark in the toolbar. For the moment, just follow the indicated steps. Click into the appropriate fields on the left side of the window to reproduce the selections as shown in the picture below.

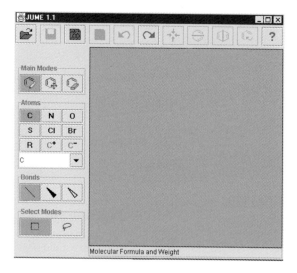

With the left mouse button click into the drawing field. A methane molecule is formed and you see the following picture:

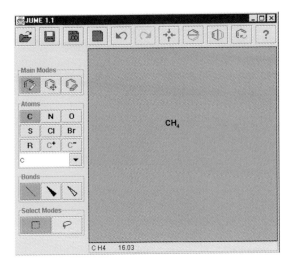

Whenever you make a mistake, you can reverse the latest few actions by clicking the "Undo" button in the toolbar (fifth from left) once or several times. You can have a look at the meaning of the toolbar symbols by moving the mouse pointer over a symbol without depressing a button: A text field will appear below the symbol.

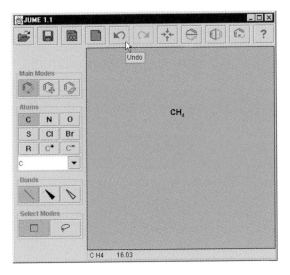

To continue drawing, move the mouse pointer over the CH4 group until it changes from black to white. Then, depress the left mouse button and keep it down: A yellow circle around the carbon atom appears.

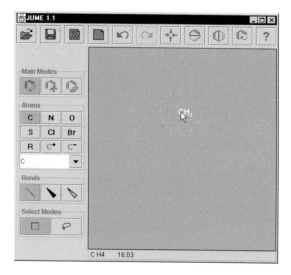

Move the mouse pointer to the right, keeping it depressed. A blue line appears connecting the starting point of the move with the current mouse pointer position.

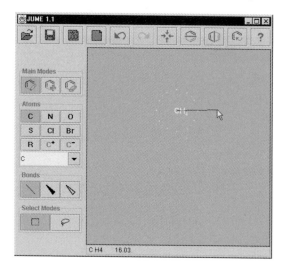

After leaving the circle, release the mouse button: An ethane molecule has been created.

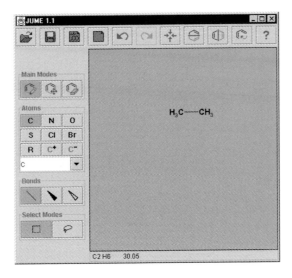

When you look at the "Atoms" field to the left of the drawing area, you see the carbon atom highlighted. Change this by clicking into the "R" field where R means a free valence. The drawing program JUME handles it exactly the same way as an atom. Move the mouse pointer over the right methyl group so that it turns white. Depress the left mouse button and move it to the right, as you did before. After releasing the mouse button, the ethyl fragment finally is formed.

To move the fragment to the main input window of Assemble, move the mouse pointer into the drawing area and depress the right mouse button. A menu pops up. Select the line "Add to Assemble Fragments".

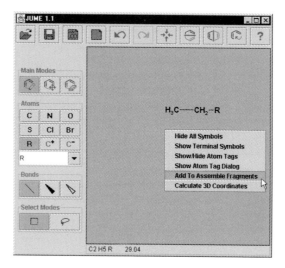

Release the mouse button. The fragment now appears in the "Fragments:" field of the Assemble main input window.

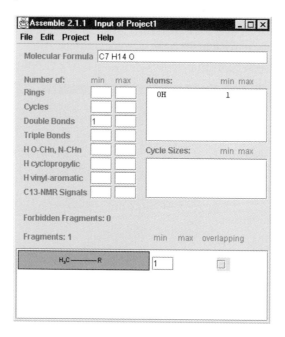

Ignore the "min", "max", and "overlapping" fields for the moment. You will learn about their meaning later. Start the generator. A total of eighty structures is generated, of which the first nine are shown as the window pops up. They combine the requested structural features, i.e., an ethyl group, an OH group, and a double bond.

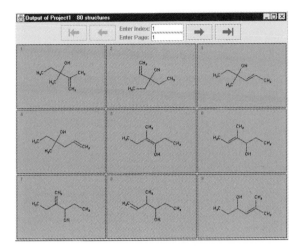

When briefly examinimg the structures, you will find that some are chemically unstable. The enol group would tautomerize to the corresponding carbonyl form. As there are too many structures generated to be handled individually, you may want to eliminate the enols. You can do this by forbidding the group. First, draw the enol as a substructure in JUME. Delete the old input by clicking the "Clear" button in the toolbar (fourth from left).

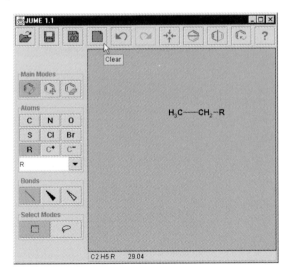

Start afresh by selecting the carbon element symbol in the "Atoms" field. Repeat the drawing steps as in the example before until you reach the ethane molecule again. You can change the bond type between the carbon atoms by clicking the bond with the left mouse button. It then turns into a double bond while the number of hydrogen atoms is adjusted appropriately. By clicking repeatedly you can circle through the sequence of single, double, triple, and unspecified bond. After forming a double bond, you get the ethylene molecule. To add the OH group, first select the oxygen element symbol in the "Atoms" field. Click onto the right carbon atom and move the mouse pointer to the right. After releasing the mouse button, an OH group has been added.

The hydrogen atoms must now be replaced by free valences except for the hydrogen of the OH group. Select the R symbol in the "Atoms" field and replace all the hydrogens by free valences. The final result may look like this:

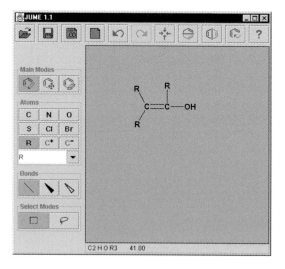

Move the substructure to the Assemble main input window as you did before. Click into the drawing field with the right mouse button and select "Add To Assemble Fragments" from the pop-up menu.

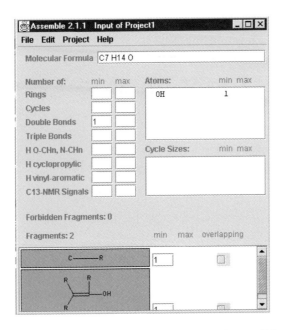

If the substructure is not completely visible, drag down the scroll bar of the window or resize the Assemble main input window by grabbing the lower border with the mouse and pulling it down. You now want to forbid the substructure. Later you will learn about the difference between nonoverlapping fragments and potentially overlapping substructure constraints. Tick the box "overlapping" and, with mouse and keyboard, change "min" and "max" from 1 to 0.

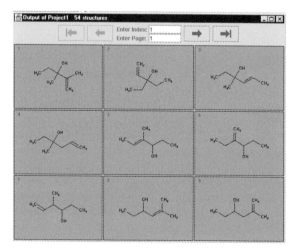

After starting the generator, 54 structures are created, a number still not easy to manage.

Certain phenomena cause the NMR chemical shifts of two or more atoms to coincide, e.g., symmetry within the molecule. Assemble can predict the number of signals observed in the broadband-decoupled ^{13}C NMR spectrum by exploring the topological symmetry of the molecule. As topological symmetry considerations are not sufficient to exactly predict the number of signals, the feature is tricky. Use it with care. You need not specify an exact number but can give a range. This allows a certain conservatism. Assume that the symmetry in the molecule under study gives rise to six signals in the spectrum. Therefore, from the seven carbon atoms in the molecular formula, two must have the same chemical shift. You can enter this information in the main input window. At the bottom of the list of features for which you can specify upper and lower limits, you find an entry "C13-NMR Signals". Set both "min" and "max" to 6 and start the generator. Thus, only two structures are created.

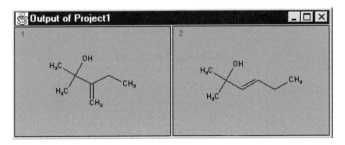

You have now reached the point where you do not need any more computational aid. You look at the structures and decide.

5.3 Fundamental Difference Between Non-overlapping Fragments and Substructure Constraints

When using Assemble, it is most important to recognize the fundamental difference between non-overlapping fragments and substructure constraints. As the molecular formula must be given, the number and kind of atoms in the molecule are known, but the kind of bonding between them is not known. Assemble generates all possible constitutions by tentatively forming bonds in a systematic way. The algorithm applies the strategy of a depth-first tree search. Bonds are formed one after the other until a constraint is violated. If this happens, the last bond is deleted and a different bond is formed. If none of the atoms has a free valence left, a complete molecule has been formed.

If there is knowledge about bonds in the molecule, the first bonds can be formed definitively, thereby reducing the size of the search tree. However, not all substructural information can be handled like this. To initially form bonds, it must be known that the structural fragments in question are nonoverlapping. Sometimes, only a single fragment can be handled like this. potentially overlapping substructures are mere constraints. Their presence, or absence, in complete candidate structures must be guaranteed by a substructure search process. While the process of initial bond formation is most effective as it reduces the size of the search tree, the retrospective substructure search is time-consuming. In addition, most of the bonds must be formed before a substructure constraint can be violated. As this happens at the periphery of the search tree, almost the entire tree has to be gone through. This behavior is an intrinsic shortcoming of the assembly process. It is, therefore, most important to carefully choose the substructural information to be used as nonoverlapping fragments. The following example illustrates the difference in performance when, in contrast to a substructure constraint, a large substructure is used as a nonoverlapping fragment.

Enter the molecular formula, $C_8H_{10}O_2$, which corresponds to four double bond equivalents. Assume that the presence of a benzene ring is known. Open the JUME drawing program and select the "Templates" button from the toolbar.

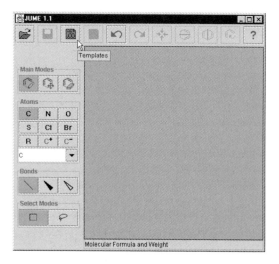

Some large, frequently used substructures are predefined, e.g., the second template is a benzene ring with all valences free. Move the cursor into the field and depress the right mouse button. Select the menu item "Copy This Structure To JUME".

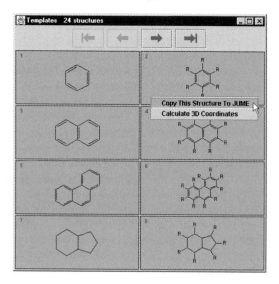

Include the substructure into the Assemble input. By default, it is considered as a nonoverlapping fragment. Run the generator and find 91 structures generated within a few seconds. Now, repeat the calculation but before, declare the fragment as a substructure constraint by ticking the corresponding box. You may not have the patience to wait till the assembly process is completed as it may take more than an hour, depending on your hardware.

By using the substructure as a nonoverlapping fragment, six bonds between carbon atoms are formed permanently. The search process starts with these bonds already formed. It takes only little time to perform the task as the remainder of the search tree is so small. In addition, no substructure has constantly to be searched for. The tree size increases more or less exponentially with the number of its levels. If the benzene ring were not used as a nonoverlapping fragment, there would be six more levels. Whenever a bond is formed, a substructure search is performed to check for the validity if the tree node. This slows down the assembly process tremendously.

5.4 Atom Tags

As it is so important to choose the nonoverlapping fragments as large as possible, some more functionality is built into the Assemble program for treating overlap. The atoms of fragments can be augmented with structural information about their immediate environment. The scope of these atom tags reaches beyond the fragment. An atom can have as many atom tags as desired. It is important to note that atom tags can only be added to nonoverlapping fragments, not to substructure constraints. If you tick the "overlapping" box, tag information is ignored.

5.4.1 Neighboring Atom Tag

Possibly, the most important atom tag is the neighboring atom tag. It specifies how many atoms of a particular element are directly bonded to the atom in question. These atoms can, but need not be part of the fragment. Optionally, the hybridization of the atoms as well as the bond type can be specified. The number of atoms may be given as a range.

Example 1

The presence of an ethoxy group can easily be derived from the ^{1}H NMR spectrum by the integrals and the coupling patterns of the signals involved. The chemical shift $\delta >$ 4 for the CH_2 group demands an oxygen atom to be bonded to the CH_2 group. The neighboring oxygen atom may be part of the fragment. However, only a single ether-type oxygen atom with a free valence must appear in all of the nonoverlapping fragments. If a second oxygen atom of this kind is included in a fragment, one must guarantee its nonidentity with the first one. This is normally not possible as no experiment will provide this information. The presence of an oxygen atom is, therefore, often indicated by a neighboring atom tag. To add a tag to an atom, first draw the fragment without tags as you learned before. Draw an ethyl group. After finishing, change the main mode of the program from drawing to selecting/moving by clicking the middle field in the "Main Modes" block, as shown below.

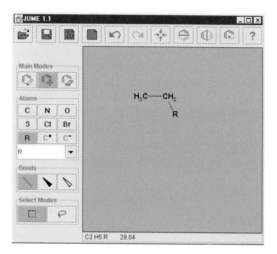

In this mode, click on any atom to select it. Its color changes to red. You can select as many atoms as you want. Deselect the atoms by clicking on an empty spot in the drawing area. If no atom is selected, clicking an empty spot will select all atoms. As you want to add an atom tag to the CH_2 group, select it by clicking on it.

Atom tags are set up in a separate window. To pop it up, move the mouse pointer anywhere into the drawing area and press the right mouse button. From the appearing menu select the row "Show Atom Tag Dialog".

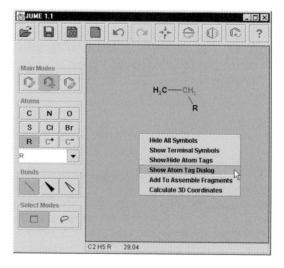

After releasing the mouse button, the atom tag dialog window pops up.

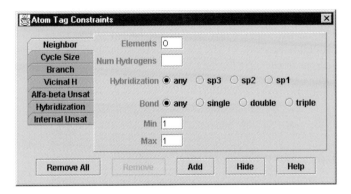

Seven kinds of atom tags can be managed. Select the neighboring atom tag by clicking the top row. Choose oxygen as the element symbol of the neighboring atom. If required, you can specify how many hydrogen atoms the neighboring atom has to be bonded to. As you do not want to specify this property, leave the field blank, or clear it if you find an entry. Choose hybridization and bond type to "any" by clicking the corresponding buttons. Finally, set both the "Min" and "Max" number of occurrences to 1. You now have the desired tags. Apply it to the previously selected CH_2 group by clicking the "Add" field. You will find the tag as a red text in a yellow field next to the tagged atom.

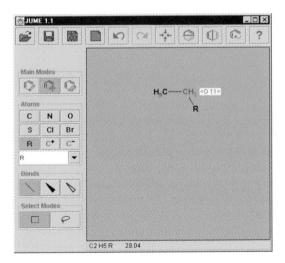

This fragment can now be moved to the Assemble main input window as you did before.

Example 2

The number of hydrogen atoms specified to be bonded to other atoms is a lower limit. In the assembly process, a CH_2 group can be expanded to a CH_3 group. If this is to be prevented, the upper limit for the number of hydrogen atoms can be specified by a neighboring atom tag.

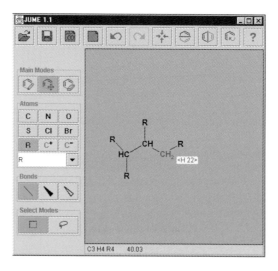

Although the tagged CH_2 group has a free valence, it will not change to a CH_3 group.

Example 3

An atom cannot be part of a fragment if its element type is not exactly known. Consider the situation that the ^{13}C chemical shift and the IR frequency can be accounted for by an amide or an ester linkage. The atoms of the carbonyl group can be part of a fragment, but not the hetero atom bonded to it, though no other oxygen or nitrogen atom may appear in the other fragments. If the element is not exactly known, the hetero atom can only be included in a neighboring atom tag. As there is already an oxygen atom bonded to the carbonyl carbon atom, the hybridization sp^3 for the hetero atom is specified to avoid interference.

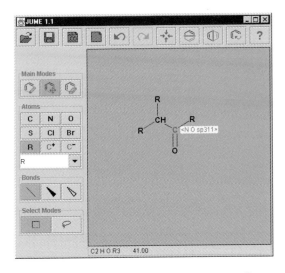

The free valence of the tagged atom has to become an sp^3-hybridized oxygen or nitrogen atom. The carbonyl oxygen atom is sp^2-hybridized, so the tag does not apply to it.

5.4.2   Cycle Size Atom Tag

Atoms in small rings often show a behavior clearly distinct from unstrained environments. If a particular atom is believed to be member of a ring, this information can be passed to Assemble by means of the cycle size atom tag. The cycle size can be given as a range. The minimum and maximum number of cycles containing the atom in question can also be specified. The feature can be used to forbid that the atom is a member of a ring by specifying the corresponding maximum occurrence as 0.

Example

A very strong band is found in the IR spectrum at 1780 cm^{-1} attributed to a carbonyl group. In the ^{13}C NMR spectrum, there is a signal at 177.8 ppm corresponding to a carbonyl environment. As this chemical shift is too low for a ketone, a hetero atom must be bonded to the carbonyl group. Unusually high IR frequencies are experienced when the carbonyl carbon atom is a member of a small ring, in this case most likely a 5-, maybe 4-membered ring. Therefore, the compound is a 4- to 5-membered lactone or lactame. The latter is excluded by the molecular formula $C_5H_8O_2$. This information is expressed by the following cycle size atom tag:

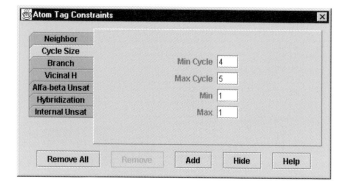

This tag information is applied to the carbonyl carbon atom of an ester linkage.

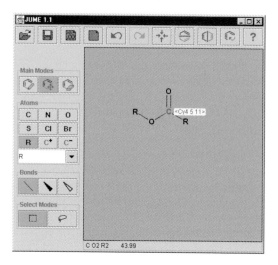

5.4.3   Branch Composition Atom Tag

Occasionally, the elemental composition of a fragment is known, but with no further information about the connectivity within the fragment. The mass spectrum may show the loss of m/z 71 from the molecular ion. This may correspond to a direct fragmentation next to a carbonyl group. The carbonyl itself accounts for m/z 28 and the remainder of 43 u would arise from C_3H_7, a propyl or isopropyl group. As C_3H_7 is a terminal fragment, irrespective of its connectivity, the information can be passed on to Assemble by the branch composition atom tag.

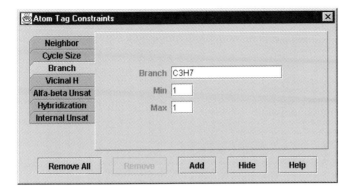

The tag is then assigned to the carbon atom of a carbonyl group.

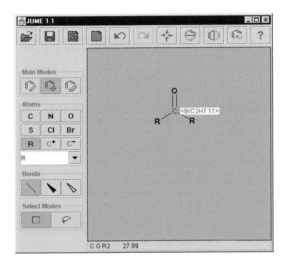

5.4.4 Vicinal Hydrogen Atom Tag

The coupling patterns in ^{1}H NMR spectra yields information about the intermediate environment of an atom. In first-order spectra, it is often possible to determine the number of vicinal hydrogens, i.e., hydrogens that are three bonds apart from the resonating hydrogen atom(s). It is not possible, though, to determine the exact connectivity. Assemble allows you to specify the number of vicinal hydrogen atoms coupling with a particular H-bearing atom. As hydrogen is treated as a special case, the information is placed on the atom that is bonded to the resonating hydrogen(s). Request exactly five vicinal H atoms by means of the vicinal hydrogen atom tag.

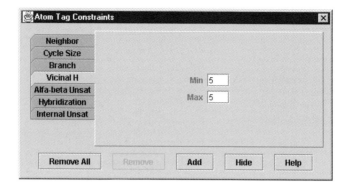

The central CH group in the two fragments shown in the window below has five vicinal hydrogens. Although the connectivity is different, they both satisfy the atom tag. In first-order spectra, assuming similar coupling constants, the signal of the central CH group would appear as a 6-line system in both cases, without the possibility of distinguishing between the two cases. Note that the content of the window is not suited to be passed on to Assemble as it contains two separate fragments. It is only shown here as illustration.

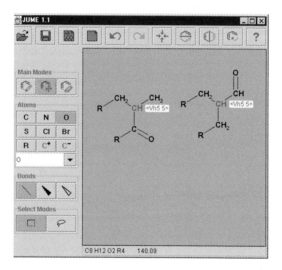

5.4.5 α,β-Unsaturation Atom Tag

Sometimes, an atom is known to be part of an unsaturated system. For example, the carbonyl group of a ketone leads to a ^{13}C NMR signal at $\delta < 200$, which suggests the ketone to be α,β-unsaturated. Assemble allows you to specify this environment. The atoms required to form the unsaturated system may, but need not, be part of the fragment.

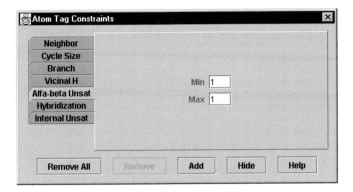

The tag is applied to the carbon atom of the carbonyl group.

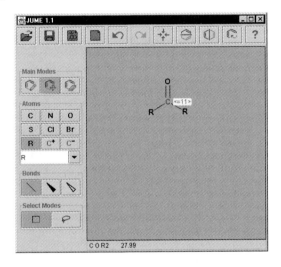

5.4.6   Hybridization Atom Tag

The hybridization atom tag can be used to specify the hybridization of an atom. It is important to remember that Assemble handles constitutions. The hybridization in this context is information about the bond types emerging from the atom, not their spatial arrangement. A nitrogen atom with three single bonds may be planar or not, it is considered sp^3-hybridized.

Example

The hybridization of a carbon atom has a great influence on its NMR chemical shift. Modern 2D NMR experiments yield connectivity information without defining any kind of bond type. Hybridization, however, which is also readily available from the chemical shift information, partially makes up for this deficiency. To declare atoms sp^3-hybridized, set up a hybridization atom tag.

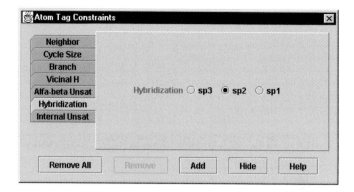

The following fragment is a typical example as derived from 2D NMR experiments. The bond types are not known, the hybridizations are.

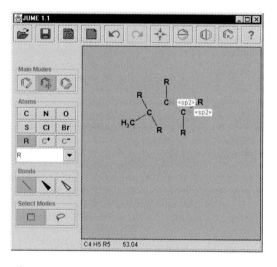

Although two sp^2-hybridized CH groups are directly bonded, it is not clear whether the bond between them is single or double, so it is left unspecified.

5.4.7 Internal Unsaturation Atom Tag

Any two atoms within a fragment are allowed to form a bond between them in the assembly process. Sometimes, it is known that, within a particular fragment, no additional bonds are allowed. This information can be passed on to Assemble by employing the internal unsaturation atom tag. The tag does not have any parameters and can be placed on any atom of the fragment. It precludes the formation of any new multiple-bond linkages within the entire tagged fragment. Cycle formation as a result of internal bridging, i.e., not involving any atoms outside the tagged fragment, is also forbidden.

Example

When a benzene ring is known to be present in the molecule, it is possibly not sufficient to specify it as a fragment. In the assembly process, two free valences of the ring could form an additional bond across it. The resulting ring system has a behavior clearly different from that of a benzene ring. Internal bonding is, therefore, forbidden by tagging any atom of the fragment.

It can be applied to any atom of the fragment or to several atoms.

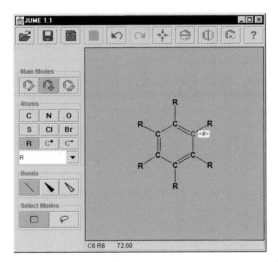

5.5 Assemble as a Tool in Structure Elucidation

One of the program's classical fields of application is the structure elucidation of organic compounds with spectroscopic methods. The following simple example may illustrate how a structure generator can enhance the productivity and reliability in structure elucidation. Five spectra of an unknown compound are given, i.e., the mass spectrum, the IR spectrum, the broadband decoupled and off-resonance decoupled ^{13}C NMR spectra, and the ^{1}H NMR spectrum.

Before Assemble can be used at all, the molecular formula must be known. This is straightforward for small molecules as in the present case. All spectra are briefly examined and only the most obvious and trivial information is interpreted.

The highest signal in the mass spectrum (EI, 70 eV) appears at m/z 114. Although the signal is weak, it is assumed to correspond to the molecular ion.

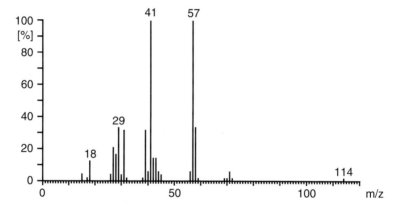

The IR spectrum (liquid film) shows a broad band from 3700–3200 cm^{-1}, which is supposed to be a trace of humidity. Obviously, the compound is hygroscopic.

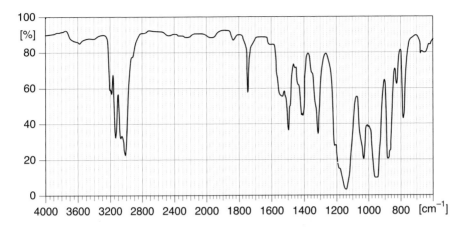

The region around 3000 cm^{-1}, characteristic of the C–H stretching vibrations, is normally not very informative since such absorptions are almost always present. The high intensity of the bands around 2800 cm^{-1} indicates the presence of hetero atoms, particularly oxygen. The only interpretable band outside the fingerprint region is the C=C stretching mode at 1650 cm^{-1} indicating the presence of a C=C double bond. The most valuable information from an IR spectrum is usually the absence of certain functional groups, when the corresponding bands are missing. In this particular case, there is no evidence of hydrogen bonded to hetero atoms and, in addition, there is no carbonyl group in the molecule.

The ^{13}C NMR spectrum (25 MHz, CDCl$_3$)) shows six signals.

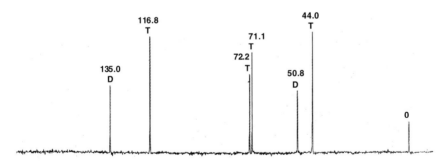

The multiplicities as available from the off-resonance decoupled spectrum are indicated as upper-case letters at the top of each signal. All carbons are bonded to hydrogen. There are two CH and four CH$_2$ groups at a minimum. The chemical shift of the triplets at δ 72–71 again indicate the presence of hetero atoms, most likely oxygen.

The ^{1}H NMR spectrum (100 MHz, CDCl$_3$) shows a number of complex signals with integral ratios of 1:2:2:1:1:1:1:1 (from left), adding up to 10 H atoms. This corresponds to the lower limit found from the ^{13}C NMR spectra.

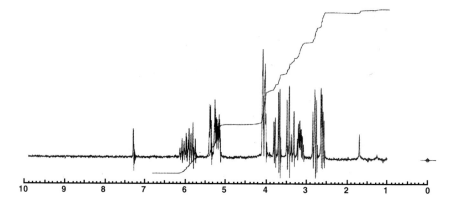

So far, we have found 6 C and 10 H atoms and indications for the presence of hetero atoms, most likely oxygen. The C_6H_{10} part of the molecular formula amounts to 82 u. If the signal at m/z 114 in the mass spectrum indeed corresponds to the molecular ion, a remainder of 32 u is to be constituted without C and H atoms. This mass difference could be accounted for by 1 S atom or 2 O atoms. Assuming the latter leads us to a molecular formula of $C_6H_{10}O_2$, which corresponds to two double bond equivalents.

Now that the molecular formula is known, Assemble can be run the first time. Possibly, a previous example is still set up. You may want to shut down the program and restart it, or you remove the old entries. To delete the substructures, move the mouse pointer into the "Fragments" field and press the right mouse button. From the pop-up menu choose "Remove All Fragments".

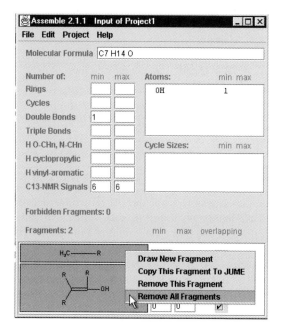

The atom constraints and cycle size constraints are removed similarly. Delete all entries directly accessible in the main window and add the new molecular formula. If nothing but the molecular formula is given, Assemble generates 4 869 constitutions, some of them looking quite odd.

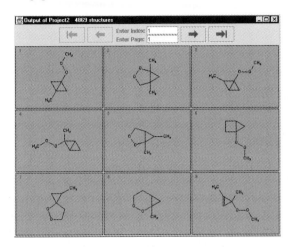

It is quite illustrative how the number of candidate structures decreases when only the most trivial information is used. The ^{13}C NMR spectra yield the number of hydrogen atoms immediately bonded to the carbon atoms. In addition, the signals at δ 135.0 and 116.8 can be assigned to sp^2-hybridized carbon atoms, while the remaining four correspond to sp^3-hybridized atoms. The information is readily entered as atom constraints.

Assemble now finds that just 26 structures are compatible with the above information. As is is a manageable number, it is worthwhile to have a closer look at these candidates. It is seen that some of them are peroxides.

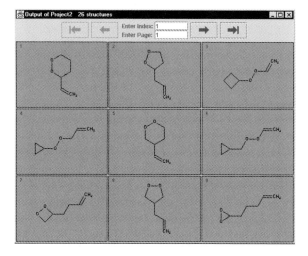

As Assemble does not consider any kind of chemical behavior, also chemically unstable molecules are generated. There is no spectroscopic evidence for the absence of a peroxide linkage. It is assumed by pure chemical intuition. To forbid the peroxides, the group is set up as a substructure constraint with the minimum and maximum limits set to zero.

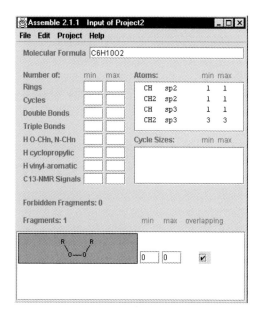

There are only 17 structures left.

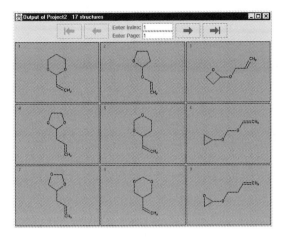

Working with Assemble adds some flexibility to the structure elucidation process. Often, the spectroscopist does not find a clue simply because of a lack of ideas. With different candidate structures available, one can attempt to find discrepancies between the spectral information at hand and that of some of the candidates. The first structure generated is obviously inconsistent with the spectra since the molecule contains a CH_2–CH_2–CH_2 linkage. The chemical shift of the center CH_2 group is expected to be $< \delta\ 2$ in the ^{1}H NMR spectrum and $< \delta\ 30$ in the ^{13}C NMR spectrum, both values contradicting the experimental findings. Some other candidates too show a CH_2 group between two carbon atoms, which is in disagreement with the spectra. It, therefore, seems reasonable to forbid such a group. It is set up as a substructure constraint.

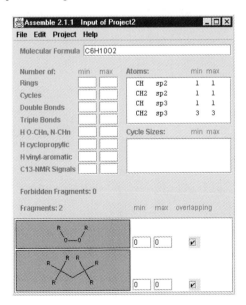

There are only 5 structures left.

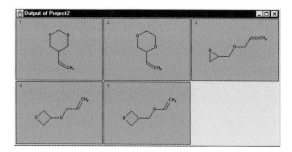

A further reduction of candidates is not worthwhile as five structures can easily be handled individually. Just for the sake of practicing, you may want to apply some more restrictions. Structure no. 5 of the output shows an oxygen atom bonded to the C=C double bond. This is in contradiction with the NMR spectra. The isolated spin-spin interaction system of the double bond would lead to a less complex signal pattern than experimentally found. In addition, the chemical shift values of the CH group next to the oxygen would be substantially higher than the observed ones in both ^{13}C and ^{1}H NMR spectra. One way to pass this additional information to Assemble is to set up the double bond as a fragment and use the neighboring atom tag for the CH group.

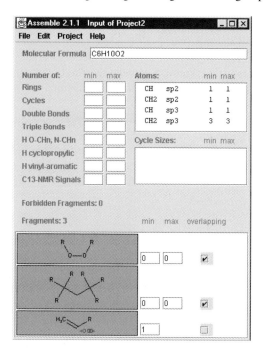

The new output no longer includes the offending structure. As none of the other candidates shows this feature, it is also the only one excluded.

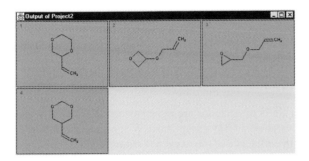

Structure no. 4 shows a CH_2 group between two oxygen atoms. The chemical shift of the corresponding signal would be expected around δ 100 in the ^{13}C NMR spectrum, ca. 30 ppm above the experimental value. Here again, as no other candidate shows this feature, forbidding the group will only remove that particular candidate. Set up the offending group as a substructure constraint and exclude it by setting the minimum and maximum occurrences to zero.

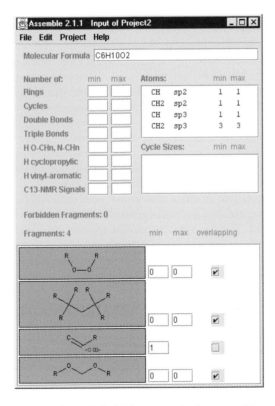

As a result, the structure is excluded. There remain three candidates.

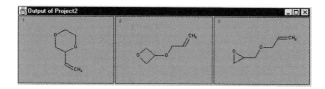

The chemical shifts of epoxides are rather special. Normally, a neighboring oxygen pushes the shift value of a CH_2 group above δ 3 in the ^1H NMR spectrum. Not so in epoxides, where the influence of the oxygen is reduced. Since in the present example, all sp^3-hybridized carbon atoms have an oxygen atom as a neighbor, the chemical shifts below δ 3 are only compatible with structure no. 3, which is indeed the correct constitution. It is the same as in Problem 5 (Chapter 3.5). It is, obviously, possible to solve the problem with spectra dating from about 30 years ago, but it is not easily accomplished without the aid of a structure generator.

5.6 Ranking According to NMR Spectral Information

Assemble accepts structural, not spectral information. The user must, therefore, interpret the spectra. This does not mean that all the interpretation has to be done manually. There are tools available to automatically transform spectral into structural information. Two of them have been integrated into Assemble. Such interpretation aids can be of great value, but their strength also bears a certain danger of misinterpretation. There is almost always some doubt as to what a spectral feature means in terms of connectivity between atoms. Occasionally, also experts make mistakes and automatic procedures are not more reliable. The user still is responsible for the correctness of the predictions. So, it is recommended to use the tools with care.

The two interpretation modules that come with Assemble estimate the NMR chemical shifts of the candidate structures generated. The values are then compared with the experimental ones. The structures are ranked according to a criterion of agreement between estimated and experimental values. Experience shows that the correct structure is often near the top of the ranking list, even if only the molecular formula and the most reliable data is provided.

5.6.1   Ranking According to the ^{13}C NMR Spectrum

The ^{13}C NMR spectral information is given as chemical shift of the signals and their multiplicity in the off-resonance decoupled spectrum. This is the information about the number of immediately bonded hydrogen atoms and is also accessible from various multi-pulse NMR experiments such as DEPT.

Try the module by using the structure elucidation example (cf. Chapter 5.5), this time providing only the molecular formula and the ^{13}C NMR data. The ^{13}C NMR spectrum shows six signals. Their multiplicity is also given: The letters D and T below the chemical shift values mean doublet (of a CH group) and triplet (of a CH_2 group).

First, run Assemble entering the molecular formula only. If you have a current project running, delete all input fields or open a new project. Specify the molecular formula as $C_6H_{10}O_2$. and run the generator. There are 4 869 structures generated.

From the "Project" pull-down menu select "Rank Output".

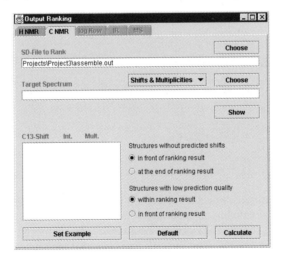

A window pops open. Choose the "C NMR" tab from the top of the window.

The entire spectrum in JCAMP format can be analyzed to provide the chemical shift information. Alternatively, the numerical values can be entered manually. The format is shown after clicking the "Set Example" button. Now, you can type in your own data.

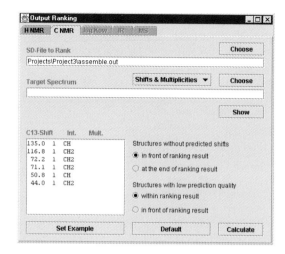

Click the "Calculate" button. Depending on the type of your hardware, the calculation may take a few seconds up to several minutes. The result is shown in a window very similar to the usual Assemble output window. It may be necessary to enlarge it by grabbing the right border and pulling it to the right.

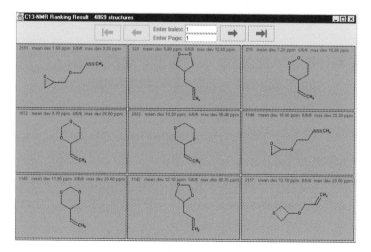

As you see, the correct structure ranks first, with a mean deviation of 1.60 ppm. The maximum deviation amounts to 3.30 ppm. This is very moderate and is considered typical for a correct structure. Evidently, these deviations are much better than those of the structures ranking below. This strongly indicates that the first structure represents the correct solution.

The numbers 6/6/6 at the top of each isomer mean, respectively, that six shifts (i.e., in the present example, all of them) have been assigned to carbon atoms, that the structure is predicted to have six signals in the spectrum, and that six experimental shifts have been specified.

5.6.2 Ranking According to the ¹H NMR Spectrum

Switch to the "H NMR" method and click the "Set Example" button as you did before with the ¹³C NMR data.

The ¹H NMR spectral information is given as chemical shift of the signals and their integral. In addition, a coupling pattern may be specified. If you do not want to provide this information, enter the letter m (for multiplet). Start the calculation by clicking the "Calculate" button.

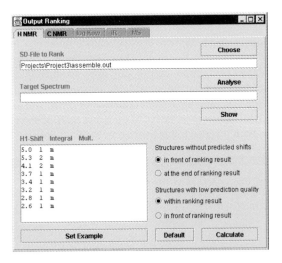

The window popping up shows the ranking results.

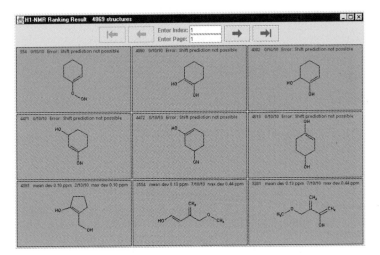

The structures look strange. In addition, the text, in red, in the output fields of the first six isomers shown says "Error: Shift prediction not possible" and when looking at the first field, you also see "0/10/10". This means that not a single chemical shift has been mapped to any hydrogen atom in the structure. All values are outside the acceptable shift interval for the particular combinations of chemical environment and signal integral. It, therefore, must be a very improbable candidate. Why does it rank first? The answer is: as a measure of security. Sometimes, the shift values of the correct structure cannot be estimated adequately owing to insufficient information about its chemical environments. If this happens, you are very likely to miss the correct solution. This is the worst possible outcome. Therefore, these structures are shown first to alert you.

The structure ranking as 7th is the first one for which no error is reported. Two of the experimental shifts are mapped. As there is ample choice for so few mappings, the mean and maximum deviations tend to be small. Therefore, these structures rank very high, though the bad mapping again shows the structure to be an unlikely candidate.

When you look at page 3 of the output with the structures ranking from 19th to 27th, you see the correct solution ranking as 26th. The results are shown as text in blue color. The mean and maximum deviations are 0.20 ppm and 0.59 ppm, respectively. All the chemical shifts have been mapped to the hydrogens of the molecule.

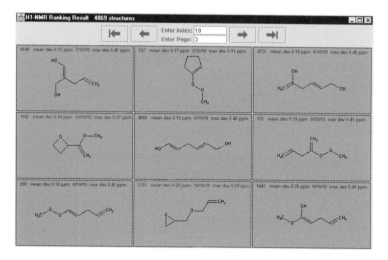

The color of the text gives you an indication about the quality of the spectrum estimation, as follows:

red: Estimation of at least one chemical shift was not possible because information about the chemical environment is not available. The corresponding shift is not used in the comparison with the experimental spectrum. Keep in mind that generated candidate structures may look strange. This is highly probable if a red entry is encountered.

magenta: The estimation of at least one chemical shift was not possible because the corresponding chemical environment leads to a wide range of the shift value. This is the case for hydrogen bonded to hetero atoms. If some less vital information about the chemical environment is missing, the estimation is of lower quality but still used for comparison with the experimental spectrum.

blue: All chemical shifts could be estimated satisfactorily.

The first entry written in blue is the structure considered to be ranking first. However, in the present example, it is not the correct one. Mean and maximum deviations are very similar to those for the correct structure. To see the next entry written in blue, you have to go forward to the structure ranking as the 67th.

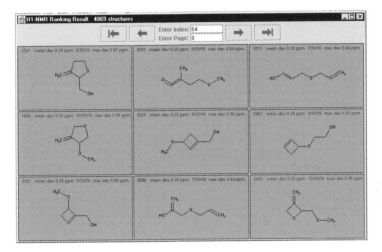

The mean and maximum deviations are still quite reasonable, which is also the case for the next few entries with all shifts mapped. Therefore, the ranking result is not as striking as when using the ^{13}C NMR data.

If you do not want the structures without predicted shifts to appear at the beginning, you can request them to come at the end of the list. Select the radio button for "at the end of ranking result" as shown in the corresponding window.

Repeat the calculation. The correct structure now ranks as the 22nd.

5.7 Postprocessing

Occasionally, the available structural information is of very diverse quality. Some information, such as the number of signals in the ^{13}C NMR spectrum, is almost without any uncertainty. Some other information, e.g., the presence of an SO_2 group according to the IR spectrum, is much less certain. In such a situation, one could run Assemble several times, trying all alternatives. However, this may become cumbersome as the execution time for each run could be substantial. A much better way is to enter only the high-quality information in a first run, thereby, of course, generating a large number of candidates. In subsequent runs, the list of constitutions is pruned by entering the information of inferior quality, resulting in shorter execution times. Postprocessing can be nested to any depth.

As an example, you may run Assemble with the molecular formula $C_6H_{10}O_2$ as before. Again, there are 4 869 structures generated. Now, select the "Postprocess Output" item from the "Project" pull-down menu.

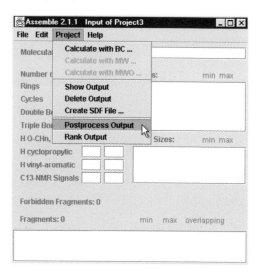

The sub-project window pops up. In postprocessing, you cannot change the molecular formula. Therefore, the corresponding input field is not shown.

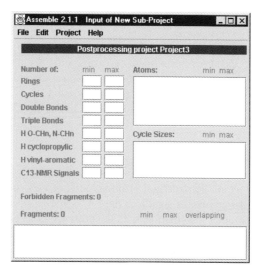

Just for curiosity, you may want to find out whether there are molecules so symmetric as to exhibit only two signals in the ^{13}C NMR spectrum. Enter 2 in the corresponding "min" and "max" fields and start a new Assemble run. You are asked to create a new sub-project. Give it a name and click into the "OK" field.

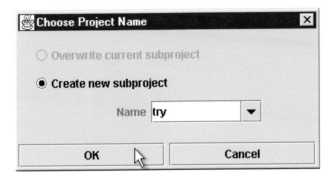

There are, indeed, six structures selected. To convince yourself that the third structure has two kinds of carbon atom environments, you may want to look at a three-dimensional picture of the molecule. Move the mouse pointer over the structure and click with the right mouse button. Select the "Calculate 3D Coordinates" menu item.

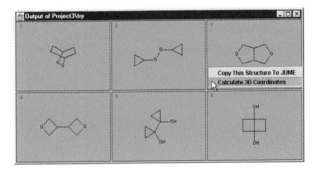

The structure appears in a separate window. For better clarity, hydrogen atoms are not shown. Move the mouse pointer over the structure and keep one of the buttons depressed while moving the mouse. The structure rotates according to your mouse movements. There are indeed two different carbon atom environments.

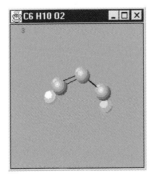

6   Structures of Compounds

Problem	Structural Formula	Name	Page
1	$\begin{array}{l}CH_3\\ |\\ CH-NH_2\\ |\\ CH_2-OH\end{array}$	(S)-(+)-2-Amino-1-propanol	19
2		(S)-β-Methyl-γ-butyrolactone	29
3		(±)-1-Phenyl-1,2-ethanediol	37
4		1,1-Bis(4-chlorophenyl)-2,2,2-trichloroethane (DDT)	47
5		Allyl 2,3-epoxypropyl ether	55
6		Methacrylic acid cyclohexyl ester	67

Problem	Structural Formula	Name	Page
7		3,4-(Methylenedioxy)-phenylmethanol	77
8		Vanillin; Isovanillin	85
9		2-Morpholino-2-phenylacetonitrile	97
10		(S)-(–)-Nicotine	107
11		4,4,4-Trifluoro-1-(2-thienyl)-1,3-butanedione	121
12		Piperine	133
13		Phosphonoacetic acid triethyl ester	143

Problem	Structural Formula	Name	Page
14		Dimedone	151
15		Benzilic acid 2-(dimethylamino)ethyl ester hydrochloride	161
16		α-D(+)-Glucose pentaacetate	171
17		3-(3-Indolyl)propionic acid	181
18		(*E*)-2-[(4-Methoxy)ben-zylidene]-3,4-dihydro-1(2*H*)-naphthalenone	191

Index